Radio Society of Great Britain

30130500063939

Published by the Radio Society of Great Britain, Cranborne Road, Potters Bar, Herts EN6 3JE.

First published 2003.



ISBN 1 872309 94 1

*Publisher's note*



Cover design: Braden Threadgold, Potters Bar.
Illustrations: Ray Eckersley and Bob Ryan.
Subediting and typography: Ray Eckersley.
Production: Mark Allgar, M1MPA.

Printed in Great Britain by Nuffield Press, Abingdon, Oxfordshire.

Extra resources for this book are at www.rsgb.org/books/extra/command.htm

# Contents

# Preface

omputers and microcontrollers cover a very wide field of devices and systems, from the small low-cost, single-chip device used to control domestic appliances such as toasters and washing machines, right up to the mainframe computers used by large organisations. The subject of digital signal processing pervades many areas of amateur radio now from simple audio processing to the latest data modes. Within this book we will cover the typical range of devices that are available to the home user and constructor, and see how they can be made to work in conjunction with many amateur radio applications to provide operator assistance, give access to new communications techniques and add a level of sophistication and usefulness to a whole range of home-constructed projects that would have been impossible, or at least very complex, without involving a processor.

In Chapter 1, we look at the uses that can be made of the personal computer. Some of the general-purpose applications software that is of immediate use to the radio amateur will be covered, such as spreadsheets and how this is now the software of choice for exploiting measurement data and performing calculations that were once the preserve of custom software. An introduction to computer programming shows how to write simple programmes and how these can be built up into quite complex software suites. Data modes are covered in outline, although this aspect of using computers for on-air communications is covered very comprehensively by Murray Greenman, ZL1BPU, in his book listed in the references for this chapter, and so will we will not go into too much detail here.

Chapter 2 expands the uses that can be made of the PC by interfacing to the outside world through its various input/output ports and covers these in some detail, including the software and hardware needed to make use of the parallel and serial (COM) ports. The design of hardware to plug into the older PC busses as well as timing issues, the use of other ports, isolation and EMC issues are discussed.

Chapter 3 introduces the PC as a tool for controlling amateur radio transceivers. Software techniques covered include band monitoring, spectrum analysis and spectral usage as well as automatic beacon monitoring. The intricacies of control of the transceivers supplied by different manufacturers is described, with various software techniques to make this aspect easier to handle. The CFAR technique for signal detection in noise is described, and also how the Global Positioning Satellite system offers new possibilities for signalling with its very accurate timing.

Chapter 4 introduces the microcontroller, concentrating mainly on the PIC device from Arizona Microchip, but also covering the products offered by some other manufacturers. A general introduction to programming the PIC is augmented by some of the basic routines common to many of the uses to which these devices may be put in amateur radio applications – such as serial interfacing, analogue-to-digital conversion, displays and basic user input/output devices. Chapter 5 introduces a number of simple stand-alone projects making use of the PIC, such as applications for automatic SWR measurement, frequency measurement, data logging, and a low-cost direct digital synthesiser.

Chapter 6 covers the area of remote control in some depth. Software and hardware in this section is described not so much for it to be used directly in complete projects, but as a source of programming techniques and algorithms that users can draw on and adapt at will to their own designs. Hardware for interfacing RF and telephone line signalling is described, along with some security and coding issues. PIC techniques are expanded in Chapter 7 where a number of projects are covered that involve more complex and subtle programming, and hardware tricks.

The principles behind digital signal processing are introduced in Chapter 8, with many of the basic algorithms common to most DSP applications described. Chapter 9 covers hardware and systems needed to use DSP at home and covers the specialities of DSP chips, their starter kits and evaluation modules.

Finally, in Chapter 10 we are indebted to Alberto de Bene, I2PHD, for a comprehensive description of Windows programming and using the sound card. Alberto describes how using this operating system introduces new concepts in programming, and takes a digital comb filter programme as a worked example of how software for the sound card is developed in Windows.

## Software listings

Now a note about software and programme listings. A lot of this book is taken up with describing various software techniques and algorithms. In many cases these will be shown as a listing of the commands needed to fulfil the tasks described, although most of these listings cannot be used in isolation. All computers and microcontroller programmes require some overhead to set up internal registers, input/output ports and suchlike. In many cases this overhead has a lot in common between multiple programmes, although the actual names of registers and ports will likely be different. Due to the general similarity, this overhead code will not be shown for the majority of listings. The techniques needed to generate the specific starting code will be covered in the introductions in each case, and usually can be adopted with slight modifications as appropriate to suit any end application. Where listings are given for various techniques and algorithms, each listing will need to be included with all the additional set-up instructions specific to the way it is being used; all listings will have sufficient comments to allow this to be generated where the information is not self-evident. More comprehensive listings will usually be available from the RSGB website page that accompanies this book, including many stand-alone programmes that are not shown in full here. The page is at:

**http://www.rsgb.org/books/extra/command.htm**

In the chapters on DSP, a number of the algorithms or DSP routines are illustrated by what I (and a college lecturer from the dim and distant past) call *pseudocode*. This is a fictional computer language making use of a reasonably straightforward set of basic commands that are self-explanatory, and with obviously named variables. It often looks a lot like the Basic language, but introduces and manufactures various functions as required. This route has been chosen as a number of readers will want to implement the DSP routines on Windows-based computers using their own preferred language. All the overhead commands associated with Windows programming would only obfuscate the basic underlying code, which is many cases is straightforward to convert to another language. Also, conversion to native code for various DSP chips is made simpler if there are no extraneous commands in the original listing.

*Andy Talbot, G4JNT*


## Acknowledgements

Alberto de Bene, I2PHD, for Chapter 10 on Windows programming – perhaps I will yet take the mental plunge to go this way too!

Peter Martinez, G3PLX, for introducing me to the Motorola 56002EVM and providing much help over the years as I was learning about DSP. Also for virtually re-drafting Chapter 8 after proofreading the first attempt.

Paul Phillips, G4KZY, for the section introducing the AVR microcontroller, the design of the HF-08 data logger hardware and a number of penetrating thoughts over the years.

Lee Wiltshire, G0IAY, for introducing me to PICs back in the dark ages, and providing the first few lines of code that still start many of my programmes to this day.

Mike Porteous and Barry Sumner, two work colleagues, for several heated discussions on DSP algorithms and techniques around the office whiteboard.

I would also like to acknowledge the many folk who gave permission to use their material, pictures and contents of their websites.

# CHAPTER 1

# The personal computer

## The beginnings of home computing

Although computers as we know them have existed in some form or another for over half a century, they only began to appear in amateur radio in the 'seventies. Back then, computers were nearly always home constructed around one of the few microprocessor chips available, and programmed, usually from first principles, by the builder. Programming the very first models had to be done in the native machine code of the microprocessor, frequently by setting the code to be programmed on DIP switches then pressing a programme button, one byte at a time. Needless to say, large programmes could take many days or weeks to get going, as just one switch set erroneously would crash the whole computer. Kits and ready-made boards with ready-installed simple operating systems and machine code assemblers soon became available, and these made programming a lot easier for many users. It was about this time that radio amateurs started introducing computers to radio, and probably one of the first uses was to replace the noisy mechanical teleprinter for radio teletype (RTTY) operating.

Interest in home computing grew rapidly, and complete ready-made home computers with proper keyboards, usually making use of a domestic television as the monitor, soon became available. These were initially eight-bit machines based around standard microprocessors, and most domestic home computers came with a high-level programming language to give the machine a wider appeal than just to computing enthusiasts. The language supplied was nearly always a version of Basic (Beginners All-purpose Symbolic Instruction Code) which still exists, in a number of different variants, to this day. The BBC Computer which appeared around 1980 was arguably the stimulus for a huge increase in computer home usage. This was an eight-bit machine, supplied with a very comprehensive version of Basic, and had huge flexibility for expansion and interfacing to the outside world. The ease of interfacing meant that the BBC Computer was still in use in education as well as at home until well into the 'nineties.

These early eight-bit machines were rarely powerful or fast enough when programmed in Basic to work directly with real-time data or audio from an amateur radio system, and the rare few who did want to pursue this route were forced to go back and learn the intricacies of machine code or

> ### Processing power in bits
>
> The 'size' of a processor or microcontroller is defined by the number of bits used to form its instruction cycle or data word. The very earliest microprocessors used only four bits, allowing $2^4 = 16$ different values. These chips were of very limited use, being restricted to jobs such as controlling washing machines! Very soon eight-bit processors such as the 6800, 6502 and Z80 chips became available and these formed the basis of the first home computers. 16-bit processors followed and now the home PC with a 80386 or higher processor makes use of a 32-bit architecture.
>
> The number of bits defines how numbers are handled. An eight-bit machine can only handle the integers 0 to 255 in a single word. By making use of two words, integers from 0 to 65535 (or −32768 to 32767) are possible. Floating point numbers as used for most mathematics require at least 32 bits, so eight-bit and 16-bit machines have to use two or four words per value, making them inherently slower as multiple memory locations need to be accessed for each value.

assembler. Hence the main use to which home computers were put was for running applications software. They were also connected to radios for processing data to and from an interface that converted the audio signals carried by the radio link to data – this interface was the *modem*, short for 'modulator/demodulator'. This configuration made possible first radio teletype modes, then more advanced communications modes incorporating error correction such as AMTOR, introduced to the amateur world by G3PLX.

During the mid-'eighties packet radio also became popular, usually making use of a dedicated terminal node controller which contained a separate microcontroller in its own right, although some packet radio software could run completely on a host computer and required just a simple one-chip modem to interface to the radio. Once such system was produced by Baycom [1] and is probably the simplest way of getting a basic packet radio station together.

The IBM Personal Computer first appeared around 1983 and was at first used mainly for business due to its cost. However, the popularity of this model rapidly grew, bringing price reductions as more and more manufacturers produced 'clones'. The PC started out initially as an eight-bit machine but soon the increased capability of 16-bit processing came with the introduction of the 80286 processor models.

Now, home and business use was driving the market, making the PC and its Microsoft MSDOS operating system dominant throughout the world. Microsoft introduced its graphical user interface-based Windows operating system in 1991. Shortly afterwards, processor enhancements meant that 32-bit computing power became the norm, and the Windows operating system was enhanced until we reach the present situation where many families have a 32-bit computer in their homes. During this rapid development in the lifecycle of the personal computer, there was a parallel massive software development programme with much of the later software requiring an upgrade to a more powerful machine to make use of its full capabilities.

This meant that older, slower machines became obsolete before they might otherwise have done, and appear frequently as surplus at low cost.

Remember that a lot of very useful software can run perfectly happily on machines 15 years old – considered antique by modern standards! Such machines are often available as scrap and in some cases, particularly for interfacing and control, can be *more* useful to the radio amateur than the latest 32-bit machine running Windows.

# Computer operating systems

All but the very first PCs used a disc-based operating system, usually MSDOS or PCDOS. This is usually resident on a hard disc (although the OS can run from a floppy disc) and presents the user with a text-based means of communication with the computer, by typing in commands in longhand. Extensive commands are available for formatting discs, copying, renaming and deleting files, and for setting the operating parameters of the machine. This straightforward user interface hides the real complexities of the DOS operating system, however. Inside the DOS software is a whole array of mini routines for interfacing to the machine hardware and providing the various 'hooks' so that higher-level software, such as Basic or any other programming languages, can interface to the outside world in a straightforward manner.

For serious programmers who want to access more of the capabilities on a typical PC than are available through the high-level programming language, it is possible to call the numerous DOS routines directly, giving access to the whole machine's capabilities. For example, to use the mouse in a DOS environment requires separate calls to the operating system to get the current cursor X and Y co-ordinates, detect buttons pressed and show or hide the cursor, or limit its travel. The DOS operating system is called by loading processor registers and calling an interrupt.

## The Windows operating systems

The first graphical-based PC operating systems, up to Windows 3.x, still relied partially on the underlying DOS and its interrupt calls, although adding levels of complexity of its own. When running Windows 3.11, for example, it is possible to call up a DOS window which behaves exactly the same as if DOS were the native operating system in use. The later 32-bit operating systems, starting with Windows 95, then Windows 98, 2000, ME etc no longer interfaced to the machine hardware via DOS. In fact 'plug-and-play' had by now appeared so that hardware and accessories could be identified directly by the operating system, making it much simpler for the user to install new hardware.

This gives us a problem if we want to run older 'legacy' software written for DOS, and do some 'out-of-the-ordinary' tricks with a PC that could involve trying to interface it to the outside world. All is not lost, fortunately. The operating systems designed primarily for single users, such as Windows 95, 98, 2000 and ME, all have a Command Prompt utility which opens a window offering nearly all of the capabilities of the original DOS, by emulating the original operating system. Not all functions are fully compatible, however; for example, there are frequently problems with timing and delays. These aspects will be covered in more detail in the next chapter. Operating systems designed for multiple users, such as Windows NT, which are more security conscious, do not offer full DOS compatibility, and direct access to the hardware, even via what appear to be DOS interrupt calls, is not available.

## Linux

Linux is free software. But, just because it's free, doesn't necessarily mean it's free. Think 'free' as in 'free speech', not 'free beer'. In a nutshell, software that is free as in speech, like Linux, is distributed along with its source code so that anyone who receives it is free to make changes and redistribute it. So, not only is it permitted to make copies of Linux and give them to your friends, it's also fine to tweak a few lines of the source code while you're at it – as long as you also freely provide your modified source code to everyone else. To learn more about free software and the major software licence it is distributed under, called the *General Public License* (GPL), look at the Linux website. In addition to the GPL, there are many other software licences that allow you to modify the source code. The Open Source Initiative approves these licences and keeps a current list of them.

Linux is not owned by anyone. One misconception many first-time www.linux.com readers have is that this site is similar to www.microsoft.com, which is owned and controlled by the company that produces the Windows operating system.

Not so. No one company or individual 'owns' Linux, which was developed, and is still being improved, by thousands of corporate-supported and volunteer programmers all over the world. Not even Linus Torvalds, who started the Linux ball rolling in 1991, 'owns' Linux.

(However, the trademark 'Linux' is owned by Linus Torvalds, so if you call something 'Linux' it had better be Linux, not something else.)

*How to get Linux*

When you 'get Linux' you are usually getting a Linux distribution that contains not only the basic Linux operating system, but also programs that enhance it in many ways. Anyone who wants to put together their own Linux distribution is free to do so, and there are more than 200 different Linux distributions that fill special niche purposes. But new users are advised to stick with one of the five or six most popular general-purpose Linux distributions until they know a little about what Linux can and can't do.

Linux is available from a number of online software repositories, including the official websites for each distribution. For example, at www.linux-mandrake.com you'll find the Mandrake distribution; at www.redhat.com you'll find Red Hat Linux.

It helps to have a fast connection and a CD writer so you can quickly download an ISO image of the distribution and write it onto a CD. You then can load the bootable installation programs that lead you, step by step, through the process of getting Linux on your computer.

If you don't have a CD writer, you'll be better off if you buy a CD pre-loaded with the distribution (or distributions) of your choice. The more popular distributions are available in many computer stores and directly from each distribution's publisher. They sell full boxed sets of CDs or DVDs that come complete with a fancy user manual and official technical support. The average price is $25 to $80 USD. The convenience of a distribution on CDs, including manuals, generally makes your first installation so much easier that it is well worth the money, and even if you pay full retail price for a Linux distribution you will still get an incredible value.

# Applications software

This can usually be taken to mean commercially produced software packages, available from computer software vendors, as well as public domain and shareware packages available for little or no cost. Public domain software is usually written by individuals in their spare time and offered free of charge to anyone who wants to use it. It is usually provided as is, with often little follow-up support or documentation. There is a huge amount of public domain software available for the radio amateur, providing facilities as widely spaced as terrain plotting for radio propagation calculations, advanced data modes for radio communications and RF design. These days the easiest way to get just about any software package is via the web. Typing a few keywords into a search engine such as Google will usually throw up enough pointers to find a site from which software can be found.

Shareware software operates on a slightly more formalised principle with regard to payment, and in turn follow-up support. Here the software is initially obtained and run free of charge. Subsequently, if the user finds it satisfactory a payment is made to the author. Often this is voluntary and based on goodwill and a 'warm feeling of support'. In many cases, after payment has been received, further enhancements to the software may be provided, such as extra help files, increased functionality or the removal of pop-up messages saying how you should really be paying for using this software!

The third type of software is, of course, the commercial package or suite which will come with full support, documentation and a guarantee. A browse around any of the high street computer outlets or computer magazines will show the vast range available for just about any specialised task.

## Spreadsheets for manipulating and plotting results

One of the more popular types of software run on many home computers is the spreadsheet, of which probably the most popular these days is Microsoft Excel, supplied as part of the Microsoft Office suite. Apart from the usual spreadsheet functions, such as the ability to make multiple calculations across rows and columns of data, Excel has a capability of reading in external data from files in a wide variety of formats. In practice, in order to import

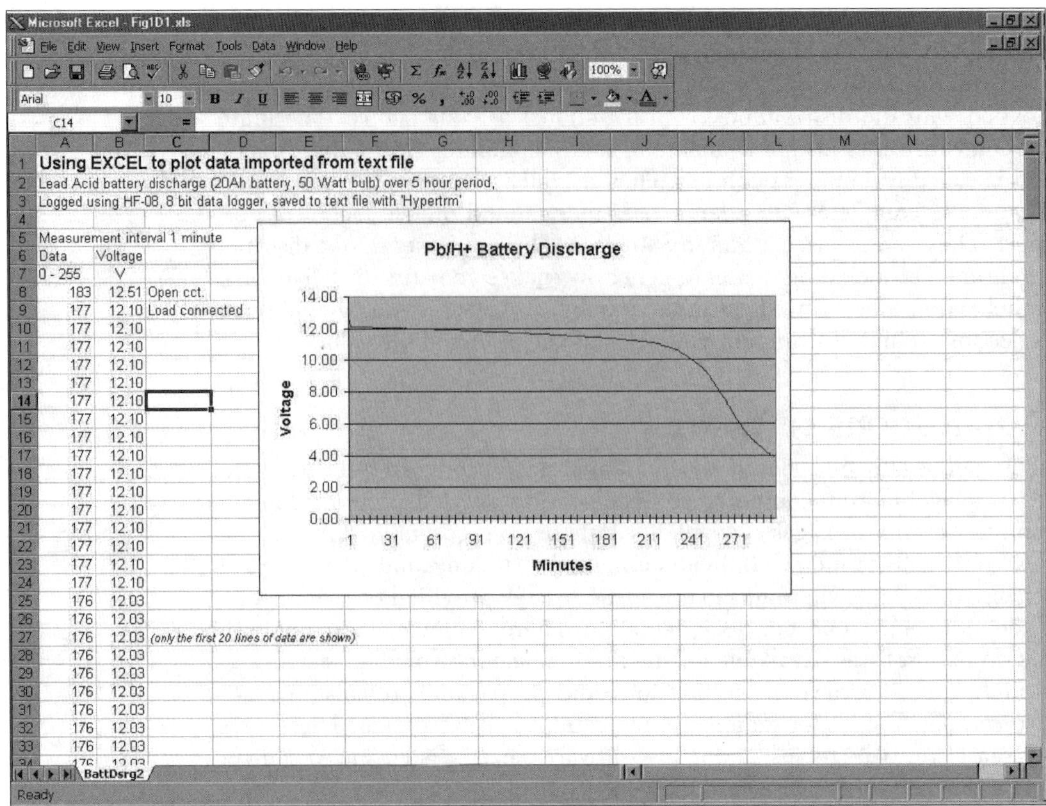



| A | B | C |
|---|---|---|
| 183 | 12.51 | Open cct. |
| 177 | 12.10 | Load connected |
| 177 | 12.10 | |
| 177 | 12.10 | |
| 177 | 12.10 | |
| 177 | 12.10 | |
| 177 | 12.10 | |
| 177 | 12.10 | |
| 177 | 12.10 | |
| 177 | 12.10 | |
| 177 | 12.10 | |
| 177 | 12.10 | |
| 177 | 12.10 | |
| 177 | 12.10 | |
| 177 | 12.10 | |
| 177 | 12.10 | |
| 176 | 12.03 | |
| 176 | 12.03 | |
| 176 | 12.03 | (only the first 20 lines of data are shown) |
| 176 | 12.03 | |
| 176 | 12.03 | |
| 176 | 12.03 | |
| 176 | 12.03 | |
| 176 | 12.03 | |
| 176 | 12.03 | |
| 176 | 12.03 | |

**Fig 1.1. Example of a graph of results, generated by reading a separately stored results file into Excel**

such data, the file containing the information to be plotted should be preferably stored as a text file. When in Excel, open this file as normal and follow the on-screen prompts with regard to format and delimiters. Once data has been read in satisfactorily, it can be processed and used to give a wide variety of graphical plots. Thus this single piece of software can be used as a back-end for displaying results from tests and measurements on just about anything we may wish. The only requirement is that the measured results, taking the form, for example, of a voltage changing over time, can be saved in a file on the computer in a form that can be subsequently read into Excel.

A suitable add-on accessory for making automatic readings of a voltage over time and saving to a file on disc is the HF-08 data logger, described in Chapter 5.

Fig 1.1 shows the spreadsheet used to plot the voltage of a fully charged 20Ah lead-acid battery powering a 50W halogen bulb. The battery voltage was divided by seven in a potential divider and fed to the HF-08 set to measure 0 to 2.5V full-scale, and output text format data each minute. Using the Hypertrm software supplied with Windows, the data from the HF-08 was saved to a .TXT file which was subsequently read into Excel. The left-hand column is the raw reading from the HF-08. With a full-scale range of 17.5V (2.5V full-scale × 7 in the potential divider) the actual battery voltage is derived from the 0 to 255 reading ($N$) supplied by the data logger as follows:

$$\text{Battery voltage} = N/256 \times 2.5 \times 7$$

This forms the second column of the spreadsheet. The graph is produced by highlighting the figures to be plotted, following the on-screen 'wizard' for producing a chart.

## Programming the PC

The other option is to write your own software. In the early days of home computers this was often the only way to use these machines, and all users became quite fluent in programming languages. These days, with the vast amount of software available, it is quite possible to manage without having to do any formal programming, other than generating spreadsheets.

However, by learning how to write your own software it is possible to unleash the maximum power, speed and performance of your computer to do just what you want, how you want it. Furthermore, there can be the satisfaction of making this available to others as public domain software, or earning some money from it by offering as shareware or selling it commercially.

To programme a computer, it is first necessary to choose a language in which to write the software. There are a limited number of languages in use these days, and each has its adherents who rigorously defend 'their' choice and state how much better it is than any other rival computer language. Needless to say, all have their advantages and disadvantages and arguably, they all provide a roughly similar performance in the finished product.

In the early days of computing the Basic language was all encompassing, with as many incompatible dialects as there were types of home computer. These days two dialects of Basic have survived the test of time, both reasonably compatible with each other. For writing simple applications for the Windows operating system, Microsoft's Visual Basic is very popular with its ability to easily generate a Windows user interface for inputting data and displaying the results. However, it is not very fast compared with other languages and is not well suited to real-time, high-speed processing or servicing external interfaces. The first versions of Basic, such as GW Basic which was supplied with DOS machines, were even slower as they operated as an interpreted language, reading, decoding and executing each line of software as it ran, rather than converting the whole programme into native machine code in the most efficient way, then running this at maximum speed.

One other surviving dialect of Basic, and the author's preferred language, is Power Basic. In fact there are two primary versions of this in use, one for writing the full-performance 32-bit Windows applications, and a 16-bit version for DOS-based operating systems, which can also run in a Command Prompt window [2]. The 16-bit Microsoft equivalent version, the predecessor to Visual Basic for DOS, is called *QuickBasic*.

<table>
<tr><td>Table 1.1. Example of a typical computer programme written in a high-level language</td></tr>
</table>

```
DO
      INPUT "Enter a number between 0 and 9" , n%
      IF n% < 0 OR n% >9 THEN
            PRINT "Error, number out of range"
            BEEP
      ELSE
            PRINT "Value entered OK"
            EXIT LOOP
      END IF
LOOP
n% = n% * 12 + 32
m = 10 * LOG10(n%)
PRINT "Result ";n%;
PRINT USING "##.#dB";m
END
```

Early versions of Basic did not encourage proper software writing practice and so this language was often sneered at by the professionals for not being capable of generating proper software and encouraging poor programming technique. The modern variants do indeed allow proper programming techniques to be used, such as local variables, variable declarations and procedures, although their use is not compulsory; older 'legacy' functions such as subroutines are still available for those who insist on using them. Early dialects of Basic required every line to be preceded by a line number but the latest versions have removed this limitation and Basic listings in this book do not show any line numbering.

Probably the widest used, and most versatile, programming language is C and its more advanced variants, C+ and C++. This is certainly the language of choice for most professional purposes and many amateurs use it too. However, programmes written in C are often more cryptic and difficult to understand (except by a C programmer) than Basic. When supported by plenty of comments and sensible variable names, Basic is a lot more self-explanatory; except possibly, to C programmers! An example of a programme written in Power Basic is shown in Table 1.1 and its C equivalent in Table 1.2.

This simple programme makes use of some of the features of Power Basic and QuickBasic. It requests the user for a number between 0 and 9, checks it for validity and, if an error has been made in entry, prompts the user to correct the error. The number is processed and an answer returned.

Basic does not require any variables to be declared, so we can jump right into the programme at the first line, DO, which sets up a repeated loop. The next line requests the user to enter a number – the text between quotes is printed to the screen telling the user what is wanted. The user now types in a value and presses Enter. The number typed in is now stored in the variable n% within the programme. The '%' means that the variable type is an short integer, ie it stores whole numbers only between the range −32768 to +32767 using 16 bits or two eight-bit memory locations to store each value.

**Table 1.2. The same programme written in the C computer language**

```c
#include <stdio.h> /* for IO functions */
#include <math.h>  /* for math routines */

#define BEEP 0x07 /* ascii character fo a beep */
#define TRUE 1
#define FALSE 0

void main(void); /* declaration of main() */

/* definition of main() */
void main(void)
{
    int n;
    float m;

    /* start inifnite loop, anything  */
    /* greater than zero is a positive test */
    do
    {
        /* read a number from console */
        scanf("%d",&n);

        /* if not in range 0 to 9 print an error */
        if((n<0) || (n>9))
        {
            printf("Error, number out of range\n%c",BEEP);
        }
        else
        {
            /* if number between 0 and 9 break loop */
            printf("Value entered OK\n");
            break;
        }
    }while(TRUE); /* always test positive */

    /* do calculations */
    /* log10() returns double precision */
    /* so cast it to a float (single precision) */
    n=n*12+32;
    m=10.0*(float)log10(n);

    /* print results %d == integer \t =tab */
    /* %3.1f = 3 sig fig with one decimal place */
    /* \n = <cr><lf> */
    printf("Result %d\t%3.1fdB\n",n,m);

}
```

The next line tests the value of n% and, if it is less than 0 or more than 9, executes the instructions listed before the ELSE statement. It first prompts the user that the value entered is out of range by printing an error message (the text enclosed between the quotes) to the screen. It then sends a beep sound to the PC speaker. If the value does not meet the test criteria, the statements between the ELSE and END IF commands are executed, here prompting the user than the value is correct, then jumping out of the DO loop to continue with the rest of the programme. The END IF statement is needed to complete the IF statement. An error will be generated when the programme is compiled and run if no corresponding END... statement is present.

The LOOP terminates the initial DO statement. LOOP UNTIL and LOOP WHILE are also possible here, giving conditional branches – their meanings should be self-explanatory. DO UNTIL and DO WHILE are also allowed, so considerable versatility is possible in setting up loop structures. The next line replaces the value of n% with a multiplied and incremented version of its original value.

The next line introduces a new variable m. Since it has no suffix it adopts the default type of a single-precision floating point number, which can take on any value between approximately $\pm 10^{-38}$ to $10^{38}$ to an accuracy of about six decimal places. It calculates 10 times the log to the base 10 of n% (its values in decibels).

The next line informs the user that the result is n% and the semicolon at the end of the line is important. It prevents the printout from dropping to the next line after printing the value of n% so that the next statement to be printed follows straight on the same line.

The next statement makes use of formatted printing. The value of m as calculated here could have any value from 10.3885 to 20.5308, but we don't need to be told the exact value, so the formatted printing statement forces the output to take on the format of two digits preceding the decimal point followed by one after it, taking into account any rounding needed. The values printed to the screen will look like 10.4dB to 20.5dB.

On reaching the last END statement the programme will terminate; in a DOS or command prompt environment it will return to the operating system. The END statement is not compulsory in Basic but allows a tidier listing.

The equivalent one-for-one C language listing is given in Table 1.2. Any text between /* ... */ are comments or explanations; they are not part of the C code and are ignored by the compiler.

Most programmes in this book written for the PC are in 16-bit Power Basic for DOS, or more usually these days for running within the Command Prompt – as that is my language of choice. In many cases they can be converted to QuickBasic with minimal changes to syntax. Conversion to the Windows versions will involve re-writing all the input and output routines to suit the graphical user interface, but the core of the software, the bit that does the real work, can usually remain as it is, with the considerable speed advantage of being able to use the full 32-bit processing capability.

It is usually quite straightforward to convert from one language into another. Frequently one-for-one translation of each line of software is possible; if not, the author making the translation will know only too well how to do the job in their own language of choice.

Alternative programming languages are Pascal, and its Windows programming variant Delphi, which are still used by some adherents; as well

## Representations of numbers within computers

Since all numerical values associated with digital machines are multiples of 2, expressing numbers in the normal decimal way is not very helpful. Instead, numbers are expressed to base 16 and the letters 'A' to 'F' are introduced to represent the values (in decimal) of 10 through to 15. By convention, such *hexadecimal* numbers are usually shown as being preceded by '0x', and this is the terminology used throughout the text within this book, it is also the format required in the C programming language. Basic, however, requires that hexadecimal numbers be preceded by '&h', and this appears in the listings.

A byte can now be expressed by using two characters, each one representing four of its bits. Some examples are:

| Decimal value | Hex representation | Binary |
|---|---|---|
| 1 | 0x01 | 00000001 |
| 12 | 0x0A | 00001100 |
| 64 | 0x20 | 01000000 |
| 103 | 0x67 | 01100111 |
| 245 | 0xF5 | 11110101 |
| 255 | 0xFF | 11111111 |

This can be logically extended to 16-bit (and higher) values :

| | | |
|---|---|---|
| 1000 | 0x03E8 | 0000001111101000 |
| 13457 | 0x3491 | 0011010010010001 |
| 49152 | 0xC000 | 1100000000000000 |

*Binary coded decimal* (BCD) is a sort of half-way house. The individual digits of a decimal number are represented in groups of four bits, each group representing the binary equivalent of its associated digit. Digits are often paired up so that a byte now represents two digits – this representation being called *packed BCD*.

| Decimal value | Hex representation | Binary |
|---|---|---|
| 23 | 0x23 | 0010 0011 |
| 127 | 0x127 | 0001 0010 0111 |

BCD representation is frequently used within microcontroller programming to simplify interfacing to keyboards and simple displays.

as older languages such as Fortran and Cobol which do show up occasionally in various older 'legacy' software.

## Variables and their representation

An area that often causes confusion amongst those new to computer programming is understanding how real-world data is stored in binary form, and the many different variable types that are available to programmers. A computer has to be able to store whatever type of user data is required for processing and this usually consists of numbers for mathematical manipulation, or text. Most data falls into one of these two categories.

All computers store data in 'words' consisting of a fixed number of binary bits per word. The word length is usually the width of the computer's data bus, as described earlier when the processing power of a computer was discussed. Word lengths in common use extend from eight bits for the very simplest microprocessors, then 16 bits for the first- and second-generation PCs, through to 32 bits for later machines, while the very latest PCs make use of a 64-bit bus and equivalent word length. Some digital signal processing chips use a 24-bit wide bus, with parts of the processing doubling this up to 48 bits.

### The byte

The most fundamental item of data seen in any computer is the *byte*, consisting of eight bits, with all the word sizes given above being simple multiples of one byte. In fact, the byte is such a fundamental concept in computing that all memory sizes and much addressing of memory is defined in terms of bytes rather than the actual word length. A 16-bit machine has a bus width of two bytes, a 32-bit machine four bytes and so on. Memory size is defined in bytes, and in practice many memory chips have a physical width of eight bits so they have to be paralleled for greater width.

### Text and characters

Letters, number and punctuation, together with print control codes such as carriage return, page feed etc, are often stored in the *American Standard Code for Information Interchange* (ASCII) representation as shown in Table 1.3. Of the 256 values possible in the eight bits making up a byte, the first 128 are allocated to all the standard control codes, numbers, letters and punctuation/procedural symbols. The remaining 128 characters are not rigorously defined and different countries have adopted some for their international characters, while other values are given over to blocks and shapes, or small drawing symbols. Within a PC the allocation of symbols to the byte values is defined in the country set-up information.

Within the PC, and most other computers, text is simply stored as a sequence of ASCII symbols, and such a set of characters is usually referred to as a *text string* – a string of characters. Where word length is greater than one byte, the characters are stored two, or more, per word. For example, a 32-bit machine would hold four symbols per word and the addressing needs to take this into account when trying to access any particular symbol. In many situations, it is necessary to know how long the string is in memory; one string may be a sentence or paragraph, for example.

There are two ways of delimiting strings. The first is to maintain a separate index table of the text, so in order to access the data the computer first looks at the index, then extracts the appropriate number of bytes from the correct memory locations. This is fine when a complex piece of software, such as a word processor, needs to continually keep track of arbitrary data, but for less-frequent text storage it is not very efficient to continually maintain a separate index table. The alternative used in many computer languages is the *null-terminated string* where the ASCII control character zero, given the descriptor NUL, is appended to the end of each string. Then to extract the complete set of bytes, the computer starts at the beginning of the data and works through it until the NUL character is detected. This process

## Table 1.3. ASCII symbols

| Decimal | Octal | Hex | Binary | Value | Notes |
|---|---|---|---|---|---|
| 000 | 000 | 000 | 00000000 | NUL | Null character |
| 001 | 001 | 001 | 00000001 | SOH | Start of Header |
| 002 | 002 | 002 | 00000010 | STX | Start of Text |
| 003 | 003 | 003 | 00000011 | ETX | End of Text |
| 004 | 004 | 004 | 00000100 | EOT | End of Transmission |
| 005 | 005 | 005 | 00000101 | ENQ | Enquiry |
| 006 | 006 | 006 | 00000110 | ACK | Acknowledgment |
| 007 | 007 | 007 | 00000111 | BEL | Bell |
| 008 | 010 | 008 | 00001000 | BS | Backspace |
| 009 | 011 | 009 | 00001001 | HT | Horizontal Tab |
| 010 | 012 | 00A | 00001010 | LF | Line Feed |
| 011 | 013 | 00B | 00001011 | VT | Vertical Tab |
| 012 | 014 | 00C | 00001100 | FF | Form Feed |
| 013 | 015 | 00D | 00001101 | CR | Carriage Return |
| 014 | 016 | 00E | 00001110 | SO | Shift Out |
| 015 | 017 | 00F | 00001111 | SI | Shift In |
| 016 | 020 | 010 | 00010000 | DLE | Data Link Escape |
| 017 | 021 | 011 | 00010001 | DC1 XON | Device Control 1 |
| 018 | 022 | 012 | 00010010 | DC2 | Device Control 2 |
| 019 | 023 | 013 | 00010011 | DC3 XOFF | Device Control 3 |
| 020 | 024 | 014 | 00010100 | DC4 | Device Control 4 |
| 021 | 025 | 015 | 00010101 | NAK | Negative Acknowledgement |
| 022 | 026 | 016 | 00010110 | SYN | Synchronous Idle |
| 023 | 027 | 017 | 00010111 | ETB | End of Trans. Block |
| 024 | 030 | 018 | 00011000 | CAN | Cancel |
| 025 | 031 | 019 | 00011001 | EM | End of Medium |
| 026 | 032 | 01A | 00011010 | SUB | Substitute |
| 027 | 033 | 01B | 00011011 | ESC | Escape |
| 028 | 034 | 01C | 00011100 | FS | File Separator |
| 029 | 035 | 01D | 00011101 | GS | Group Separator |
| 030 | 036 | 01E | 00011110 | RS | Request to SendRecord Separator |
| 031 | 037 | 01F | 00011111 | US | Unit Separator |
| 032 | 040 | 020 | 00100000 | SP | Space |
| 033 | 041 | 021 | 00100001 | ! | Exclamation mark |
| 034 | 042 | 022 | 00100010 | " | Double quote |
| 035 | 043 | 023 | 00100011 | # | Number sign |
| 036 | 044 | 024 | 00100100 | $ | Dollar sign |
| 037 | 045 | 025 | 00100101 | % | Percent |
| 038 | 046 | 026 | 00100110 | & | Ampersand |
| 039 | 047 | 027 | 00100111 | ' | Single quote |
| 040 | 050 | 028 | 00101000 |  | Left/opening parenthesis |
| 041 | 051 | 029 | 00101001 |  | right/closing parenthesis |
| 042 | 052 | 02A | 00101010 | * | Asterisk |
| 043 | 053 | 02B | 00101011 | + | Plus |
| 044 | 054 | 02C | 00101100 | , | Single quote |
| 045 | 055 | 02D | 00101101 | - | Minus or dash |
| 046 | 056 | 02E | 00101110 | . | Dot |
| 047 | 057 | 02F | 00101111 | / | Forward slash |
| 048 | 060 | 030 | 00110000 | 0 |  |
| 049 | 061 | 031 | 00110001 | 1 |  |
| 050 | 062 | 032 | 00110010 | 2 |  |
| 051 | 063 | 033 | 00110011 | 3 |  |

**Table 1.3 (continued)**

| Decimal | Octal | Hex | Binary | Value | Notes |
|---|---|---|---|---|---|
| 052 | 064 | 034 | 00110100 | 4 | |
| 053 | 065 | 035 | 00110101 | 5 | |
| 054 | 066 | 036 | 00110110 | 6 | |
| 055 | 067 | 037 | 00110111 | 7 | |
| 056 | 070 | 038 | 00111000 | 8 | |
| 057 | 071 | 039 | 00111001 | 9 | |
| 058 | 072 | 03A | 00111010 | : | Colon |
| 059 | 073 | 03B | 00111011 | ; | Semi-colon |
| 060 | 074 | 03C | 00111100 | < | Less than |
| 061 | 075 | 03D | 00111101 | = | Equal sign |
| 062 | 076 | 03E | 00111110 | > | Greater than |
| 063 | 077 | 03F | 00111111 | ? | Question mark |
| 064 | 100 | 040 | 01000000 | @ | At symbol |
| 065 | 101 | 041 | 01000001 | A | |
| 066 | 102 | 042 | 01000010 | B | |
| 067 | 103 | 043 | 01000011 | C | |
| 068 | 104 | 044 | 01000100 | D | |
| 069 | 105 | 045 | 01000101 | E | |
| 070 | 106 | 046 | 01000110 | F | |
| 071 | 107 | 047 | 01000111 | G | |
| 072 | 110 | 048 | 01001000 | H | |
| 073 | 111 | 049 | 01001001 | I | |
| 074 | 112 | 04A | 01001010 | J | |
| 075 | 113 | 04B | 01001011 | K | |
| 076 | 114 | 04C | 01001100 | L | |
| 077 | 115 | 04D | 01001101 | M | |
| 078 | 116 | 04E | 01001110 | N | |
| 079 | 117 | 04F | 01001111 | O | |
| 080 | 120 | 050 | 01010000 | P | |
| 081 | 121 | 051 | 01010001 | Q | |
| 082 | 122 | 052 | 01010010 | R | |
| 083 | 123 | 053 | 01010011 | S | |
| 084 | 124 | 054 | 01010100 | T | |
| 085 | 125 | 055 | 01010101 | U | |
| 086 | 126 | 056 | 01010110 | V | |
| 087 | 127 | 057 | 01010111 | W | |
| 088 | 130 | 058 | 01011000 | X | |
| 089 | 131 | 059 | 01011001 | Y | |
| 090 | 132 | 05A | 01011010 | Z | |
| 091 | 133 | 05B | 01011011 | [ | Left/opening bracket |
| 092 | 134 | 05C | 01011100 | \ | Back slash |
| 093 | 135 | 05D | 01011101 | ] | Right/closing bracket |
| 094 | 136 | 05E | 01011110 | ^ | Caret/cirumflex |
| 095 | 137 | 05F | 01011111 | _ | Underscore |
| 096 | 140 | 060 | 01100000 | ` | |
| 097 | 141 | 061 | 01100001 | a | |
| 098 | 142 | 062 | 01100010 | b | |
| 099 | 143 | 063 | 01100011 | c | |
| 100 | 144 | 064 | 01100100 | d | |
| 101 | 145 | 065 | 01100101 | e | |
| 102 | 146 | 066 | 01100110 | f | |
| 103 | 147 | 067 | 01100111 | g | |
| 104 | 150 | 068 | 01101000 | h | |

**Table 1.3 (continued)**

| Decimal | Octal | Hex | Binary | Value | Notes |
|---|---|---|---|---|---|
| 105 | 151 | 069 | 01101001 | i | |
| 106 | 152 | 06A | 01101010 | j | |
| 107 | 153 | 06B | 01101011 | k | |
| 108 | 154 | 06C | 01101100 | l | |
| 109 | 155 | 06D | 01101101 | m | |
| 110 | 156 | 06E | 01101110 | n | |
| 111 | 157 | 06F | 01101111 | o | |
| 112 | 160 | 070 | 01110000 | p | |
| 113 | 161 | 071 | 01110001 | q | |
| 114 | 162 | 072 | 01110010 | r | |
| 115 | 163 | 073 | 01110011 | s | |
| 116 | 164 | 074 | 01110100 | t | |
| 117 | 165 | 075 | 01110101 | u | |
| 118 | 166 | 076 | 01110110 | v | |
| 119 | 167 | 077 | 01110111 | w | |
| 120 | 170 | 078 | 01111000 | x | |
| 121 | 171 | 079 | 01111001 | y | |
| 122 | 172 | 07A | 01111010 | z | |
| 123 | 173 | 07B | 01111011 | { | Left/opening brace |
| 124 | 174 | 07C | 01111100 | \| | Vertical bar |
| 125 | 175 | 07D | 01111101 | } | Right/closing brace |
| 126 | 176 | 07E | 01111110 | ~ | Tilde |
| 127 | 177 | 07F | 01111111 | DEL | Delete |

is a bit slower as each character has to be checked as it is extracted, but is more versatile.

The Windows operating system abandons eight-bit ASCII data to give more flexibility in the number of characters that are available. Internal representation of each character is now by 16 bits, or two bytes, in a character set called *Unicode*, giving 65536 different symbols. This is more than enough to cover all international and procedural characters, with scope for future expansion.

*Representation of numbers in binary*

Within the constraints imposed by the fundamental word length defined by the processor, we need to be able to store numbers in a way that makes best use of computer memory. The types of numbers needed to be dealt with vary from simple positive numbers such as 1, 200 etc to negative numbers and fractional values. In the latter case of fractional numbers, what accuracy do we need? Is five decimal places good enough or do we need to be able to store numbers to 14 significant figures? To address these different requirements a number of different variable types are in use.

The simplest numbers to visualise are single-byte integers – the eight bits simply represent the values from 0 to 255. This variable type is not often needed – eight bit values are more usually associated with their ASCII character representation, but the BYTE variable is available in a number of computer languages.

The next is the short integer, which takes two bytes or 16 bits. This can represent all whole numbers from 0 to 65535 and the short unsigned integer (SHORT UINT) is available as a variable type in most languages. But in practice the short integer value is modified to allow for the storage and manipulation of negative numbers, and now we need to consider how negative numbers are represented.

*Negative numbers*

One method is allocate one bit as the sign bit to indicate whether the value is above or below zero. Then the other 15 bits make up the value of the number from zero to 16383. We can see that there is now an ambiguity as there are two correct values for zero, +0 and –0. Mathematics does not like this ambiguity and another way of representing negative numbers is needed.

The answer is in a system called *two's complement.* Now the allocation of bits within the word is altered slightly. The most-significant bit is allocated the value of minus its unsigned value, with the other bit allocations unchanged. So the bit values making up a 16-bit number are now:

$$-32788, +16384, +8192, +4096, +2048, +1024, +512, +256, +128, +64,$$
$$+32, +16, +8, +4, +2, +1$$

The 16-bit binary value is unchanged if the most significant bit is a zero, but if this is set the effect of adding the remaining bits to –32788 gives a negative number that can range from –32768 to –1. Some two's complement 16-bit numbers are illustrated:

```
1000 0000 0000 0011  =  -32768 + 3                  =  -32765
1110 0000 0000 0000  =  -32768 + 16384 + 8192       =  -8192
1111 0100 1010 1100  =  -32768 + 16384 + 8192 +
                        4096 + 1024 + 128 +
                        32 + 8 + 4                  =  -2900
1111 1111 1111 1111  =  -32768 + 16384 +
                        8192 ... + 2 + 1            =  -1
1111 1111 1111 0000                                 =  -16
```

From the above, it can be seen that there is only one value for zero and a positive value can be negated by subtracting one from its value, then inverting the bits. Two's complement values work automatically in addition routines, which is where the real value of this representation comes about as shown:

```
-3000    1111 0100 0100 1000
+ 100    0000 0000 0110 0100
         ___________________
-2900    1111 0100 1010 1100
```

The short integer variable type (SHORT) is one of the most commonly used variables within computer programmes, frequently used for counters and data pointers.

For larger integers the long integer (LONG) is adopted, which uses four bytes or 32 bits in two's complement representation, allowing numbers from

–2147483648 to 2147483647 to be represented. In unsigned form, values up to 4294967296 can be represented and this is the limit of address space in the PC with its 32-bit wide address bus.

QUAD integers (64 bits or eight bytes) are possible in some languages and, similarly, can be either signed or unsigned.

*Floating point numbers*

Integers are fine if we want to represent whole numbers exactly, but what about pi (3.141592653589793 etc) or –1.2?

The technique used for representation of fractional numbers is to use a binary version of the exponential notation often encountered in scientific work, eg $12345 = 1.2345 \times 10^5$ where the '1.2345' part is called the *mantissa*, and the '5' the power of ten, or the *exponent*.

32 bits are taken to represent each single-precision number (abbreviated to variable type SINGLE) and these are allocated as follows.

Bits 0 to 22 represent the mantissa, giving values that can range from 0 to a little over 8 million but normalised by dividing by $2^{23}$ so it represents a fractional value between 0 and a value just under 1.

Eight bits represent the binary exponent in two's complement format, so that it can take on any value from $2^{-128}$ to $2^{+127}$. This represents approximately the range $10^{-38}$ to $10^{38}$.

The remaining most-significant bit forms a sign bit for the result, two's complement representation not being practical for the mantissa. Thus floating point numbers can be stored that range from approximately $\pm 5.9 \times 10^{-39}$ to $\pm 6.8 \times 10^{38}$ to a precision of around six decimal places. The IEEE standard reduces this range slightly to free up bit patterns that are allocated special meanings, and the range is reduced to $\pm 3.4 \times 10^{38}$ and $\pm 1.2 \times 10^{-38}$ respectively. The freed-up codes allow certain error conditions to be stored, such as the Not A Number (NAN) situation available in many languages where incorrect programming mixes up numerical and textual data.

If a greater accuracy or a wider numerical range is desired, the double-precision (DOUBLE) type variable is available which uses eight bytes and allows numbers from $\pm 2.2 \times 10^{-308}$ to $\pm 1.8 \times 10^{308}$ to approximately 16 decimal digits of accuracy.

So we now have the variable types:

|  |  |
|---|---|
| CHAR | Character used for text etc, and usually in the form of a null-terminated string |
| SHORT | (Integer is understood) |
| UNSIGNED SHORT | |
| LONG | |
| UNSIGNED LONG | |
| QUAD | |
| SINGLE | (Floating point is understood) |
| DOUBLE | |

Confusion often reigns when converting some programmes between machines and languages, as most computer languages allow the variable type just called 'INT' or integer. Once this always meant a 16-bit integer, and LONG was used for a 32-bit integer when needed. However, with the advent of 32-bit machines running the Windows operating system, programming languages targeted at these often use the INT variable to represent one word's worth of storage, or what was once a LONG integer. This can cause catastrophic problems when transferring from one older version of a language to a later one as memory allocation of the variables gets completely lost. The safest way to avoid such problems is to avoid the INT type declaration altogether and just refer to all integer variables as SHORT or LONG. No such problems arise with floating point numbers, although the variable type FLOAT is allowed, and is equivalent to the SINGLE variable type only.

*Variable declarations*

The majority of computer languages require that all variables used in a programme be declared before they are first used. This is so that the compiler can allocate storage in memory in advance. If an attempt is made to use a variable that has not been declared, an error message informs the user of this when the programme is first run. The one exception to this is the Basic programming language where the variable type is implicit in its name; this done by appending a single character to the end of the name to define the type. For example COUNTER% and Number_Of_People% are both short integers, while Cost! and PI! are single-precision floating point. While not having to declare variables before use has its advantages when writing and debugging code, many programmers frown on the practice as it allows sloppy programming practice and allows through unwanted variables created by undetected typing errors – all modern variants of Basic now allow variables to be declared if desired.

## Crashes, recovery and data security

Everyone has heard of or had a computer crash, where some event occurs that locks up the whole machine, preventing any keyboard or mouse access. The only solution is to completely reset the computer, which loses any data that is currently being worked on. On the older PCs running DOS, it was easy to cause a machine crash as various parts of memory and input/output space being used by the operating system were fully accessible to programmers. So by accidentally accessing the wrong location, which can be done as simply as allowing an array to overflow without checking, or specifying an incorrect I/O address, a major crash could occur – most DOS crashes were not recoverable and needed a reboot.

From the Windows 95 operating system onwards this is less of a problem. Now the OS runs in what is referred to as *Protected Mode*, where it is now theoretically impossible to get at areas of memory other than those specifically allocated to the specific window or task under consideration. It is also made more difficult to directly access the computer hardware and it is this very part of the Windows OS that makes it of less value for many of our purposes since we often would like to use the I/O facilities in non-standard (and potentially dangerous) ways. It is still possible to crash a task running under Windows, but now a Ctrl-Alt-Del sequence is usually all that is needed to close the current task. Any data currently being processed is still lost, but the computer does not have to be restarted – just the task that caused the problem. Complete operating system crashes can still occur from time to time, but these are rarer and more often than not are due to older versions of operating systems being employed with modern software, or PC hardware incompatibility problems.

## Backing up

Whatever operating system is in use, there is always the possibility of loss of current data, and if the hard disc crashes or is damaged a lot of valuable data can be lost. So periodic saving and backing up of data is strongly advisable. In the event of disc crashes, there are two areas of data that need to be considered. The first is the software itself. For modern operating systems, this usually has to be completely reinstalled – often from a CD-ROM, but the installation process is straightforward enough. The second is the user data – word processed documents, for example, or personally written software which has to be saved independently on another medium such as a floppy disc, re-writable CD-ROM or a second hard disc.

However, there are precautions that, as writers of software, we can build into our programmes to make their operation survive a crash. Take as an example a piece of software that requires the user to type in a lot of data for processing – such as a contest-scoring programme. The easiest way is often to hold the data in memory as it is typed in and processed, then dump to disc at the end or whenever required. An alternative is to keep a disc file open and write each item of data to the file as it appears. This approach can be fatal – as the author once discovered when writing just such a scoring programme! All memory contents are lost when power is lost or the computer has to be rebooted following a crash, so all the data that was entered is now lost. Maintaining an open file has similar problems. Until a file is closed, some of the data that is to be written to is held in the operating system's memory, and only written to disc when this is full – furthermore the disc header information, or index, is not updated until the file is closed. So a crash while a file is open will often lead to this file being corrupted and the data not recoverable. The correct programming procedure to allow for this eventuality is to only open the file when the new data is to be written, then close it immediately afterwards. Then the most that can be lost is the one data item being processed at the time of the crash. Most contest scoring programmes are now written this way – during /P operation generators often fail and the laptop's batteries are usually dead too! There is still the remote possibility that a crash could occur during the time the file is open and, to counter even this remote eventuality, tricks could be employed such as writing to alternate files so that only one file is open at a time or making a complete file copy each time a new item is written.

It is impossible to give any general rules about how to make software as

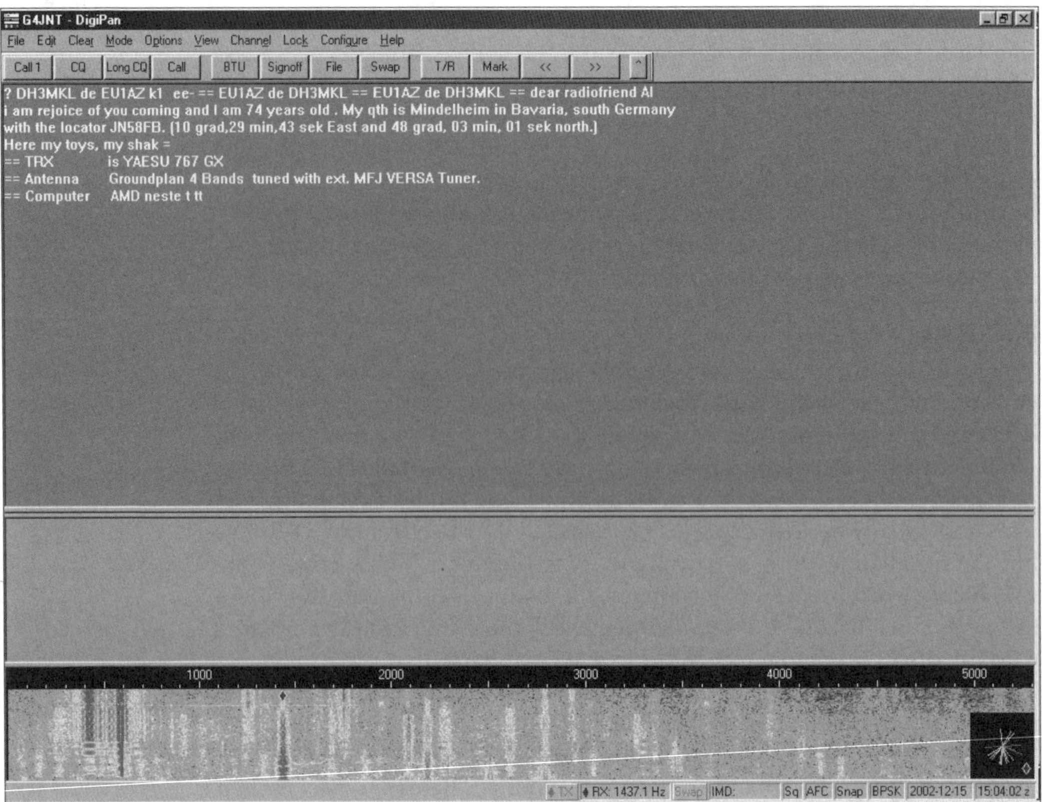


**Fig 1.2. Digipan user interface. Note the waterfall plot showing the large number of stations using PSK31 on the 14MHz band on a Sunday afternoon**

fail-safe as possible, but if a computer crash would cause excessive problems it is always worth thinking about what facilities can be built into your software to mitigate the inevitable. If it can happen, some day it will.

# Data modes

An area that has grown recently with the increased computing power now available is that of data modes. All modern PCs include a sound card, a piece of hardware that converts computer data to audio and vice-versa via high-quality digital-to-analogue (D/A) and analogue-to-digital (A/D) converters. Other features are usually offered too, such as music synthesis, but these are unlikely to be needed for any amateur radio purposes. The idea behind the modern data modes is to use the A/D and D/A conversion facilities on the sound card to allow software to work mathematically on actual samples of audio, and do the necessary processing on these to function as modems, data converters, or whatever is necessary for digital communication.

The software for this specific type of processing is usually referred to as *digital signal processing* (DSP) and is covered in more detail in Chapter 8. It is not the intention in this book to go into data modes in any detail, as these are covered comprehensively in reference [3]. The use of the sound card as

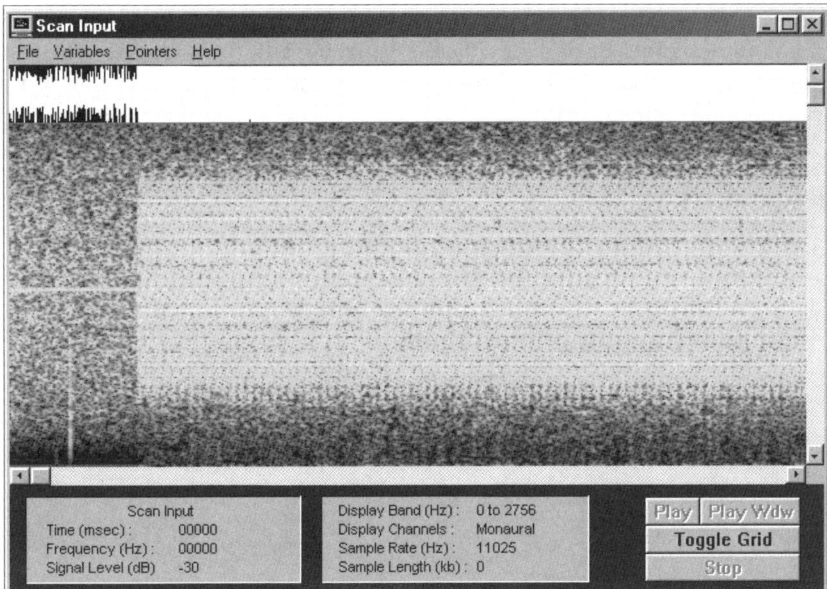


Fig 1.3. Spectrogram screen, showing a waterfall plot of a WSJT waveform

an input/output port, with circuitry to interface to external equipment, is covered in Chapter 10.

Typical software for data modes usually presents a user interface which includes a window for showing the decoded received text, another window for typing in text to be sent and a variety of controls such as sliders and buttons for setting up the correct operation of the software. Apart from the audio sound card interface, the software usually controls transmit/receive switching of a transceiver, and occasionally receiver tuning on those able to accept remote control.

Fig 1.2 shows the user interface for the DIGIPAN software written by KH6TY et al, for transmitting and receiving PSK31, a narrow-bandwidth, keyboard-to-keyboard chat mode using phase-shift keying. The mode was originally invented by G3PLX in an attempt to find a modern replacement for RTTY.

Fig 3.14 in Chapter 3 is the user screen of PCALE by G4GUO which is designed for automatic link establishment at HF, implementing the MIL Standard 188-141A signalling protocol. This software scans a number of pre-programmed HF channels for a special probe signal from specified users, looking for channels where paths may be open. It also transmits its own probe signals, and all stations taking part in the net are able to maintain a list of frequencies and the path quality between themselves so that when a communication path is required, the ideal frequency can be selected automatically.

Another type of sound card software is designed for signal analysis. Spectrogram by Richard Horne is one example of such software and is shown in Fig 1.3. Spectrogram shows the frequency spectrum of an audio input, over a user selectable bandwidth, time scale and with selectable frequency resolution. Some examples of its uses include detecting narrow-bandwidth signals buried in noise, setting up audio oscillators and test equipment, measuring signal-to-noise ratio and reading very slow CW transmissions (illustrated).

## References


[1] Baycom: www.baycom.org.
[2] Power Basic: www.powerbasic.com.
[3] *Digital Modes for all Occasions*, Murray Greenman, ZL1BPU, RSGB.
[4] Linux information: www.linux.com.

# CHAPTER 2

# Interfacing to the personal computer

nterfacing a PC to the outside world is often needed, for example, to collect data, control a transceiver or turn equipment on or off. As supplied, most home computers have very few ports with which to interface to external peripherals. The most common interface is for a printer, usually on a so-called *printer* or *parallel port*, and one or two serial ports are usually provided as a standard means to interface to a wide range of readily available peripherals such as modems, plotters, and some types of printer. Both of these ports have been available from the beginnings of the PC, and they are derived, particularly the serial port, from standards set even earlier, back in the dark ages of mainframe computers. Neither is terribly fast by modern standards.

More recently the *Universal Serial Bus* (USB) is taking over the roles of PC interfacing, and a whole range of peripheral devices now come with a USB interface. This is faster and so can be used for many more tasks – such as external disk drives or video interfaces, as well as more traditional ones such as printers. USB devices can be cascaded and hubs added, so theoretically a PC can have up to 255 external devices added.

For our purposes, probably the most straightforward interface to use for simple digital input and output from a PC is the parallel, or printer, port. It appears as a 25-pin female D-type connector on the rear panel. The hardware associated with this port is reasonably simple in concept, making it easy to drive in software. Having said that, operating systems from Windows 95 onwards can be fraught with problems in trying to gain simple access to the printer port, but for now we will assume an older-type machine running DOS.

## Interfacing using the parallel port

The printer port consists of three registers mapped into the computer's input/output area. On a 80x86-based processor as used in PCs, this is like accessing the memory area but with specific I/O commands – other processors such as the 6800 family actually do map I/O as memory. In all PCs the printer port is allocated to one of two or three specific address locations in the I/O space; most often the primary port is allocated hexadecimal

| Table 2.1 . Detailed description of LPT pin connections and polarity | | | |
|---|---|---|---|
| **Pin number** | **Signal name** | **Bit/polarity** | **Printer use** |
| 1 | −Strobe | C0 | Pulses low 0.5µs |
| 2 | Data 0 | D0 | Least significant data bit |
| 3 | Data 1 | D1 | |
| 4 | Data 2 | D2 . | |
| 5 | Data 3 | D3 | |
| 6 | Data 4 | D4 | |
| 7 | Data 5 | D5 | |
| 8 | Data 6 | D6 | |
| 9 | Data 7 | D7 | Most significant data bit |
| 10 | −Ack | S6+ | Low pulse 5µs after accept |
| 11 | +Busy | S7− | |
| 12 | +Paper End | S5+ | High for out of paper |
| 13 | +SelectIn | S4+ | High for printer selected |
| 14 | −AutoFd | C1− | Set Low to autofeed 1 line |
| 15 | −Error | S3+ | Low for error/offline/paper end |
| 16 | −Init | C2+ | Set low pulse >50µs to init |
| 17 | −Select | C3− | Set low to select printer |
| 18–25 | Ground | | |

address 0x378. Alternative locations are 0x278 and 0x3BC. The port takes up three successive locations starting at this base address, each location having an eight-bit latch associated with it so the PC can either write to or read from the port. The outputs or inputs to the latches, after suitable buffering, form the connections to the 25-pin connector of the parallel port and operate at standard TTL logic levels of nominally 0/5V.

The base location, which we will assume is the most common one of 0x378, is mapped to the eight data output lines – in normal operation this is used to send a character to the printer. On the modern bi-directional or enhanced interface, reading this location can also read the state of the same eight lines. The next location, 0x379, is used to read the status of some of the control lines and can only be used for inputting data. The third location, 0x37A, operates the strobe pulse and a few more control lines. See Table 2.1 for a full description of printer port interfacing. By making use of the commonly available printer port, eight data lines are immediately available for outputting TTL level signals, or optionally for reading eight bits of data. Four more lines can be used as outputs, provided pull-up resistors are used and note is made of which ones are inverted. Five input lines are directly accessible.

For simple control purposes this port appears ideal for interfacing to peripherals, and indeed a number of manufacturers make use of the parallel port for connecting custom hardware. Some examples are external CD writers, chip programmers and so called 'dongles' – used for security and copy protection of commercial software. The problems come with modern operating systems from Windows 95 onwards. Protected-mode operation, inherent to the 32-bit operation of these processors and operating systems, prevents direct access to the computer's hardware to protect multiple programmes running simultaneously from being corrupted. So, even when

**Table 2.2. Programme to flash a LED connected to pin 2 of the printer port**

```
def seg = &h40              'Point at BIOS information table
AddrLpt% = peeki(8)         'Table entry for address of first printer port
print "Flash a LED connected to LS Bit of the Parallel port"
print "Parallel Port at address 0x";hex$(AddrLpt%) 'Show it in hex format

do
   print "On  ";
   out AddrLpt% , 0
   delay .5            'software delay 0.5s.
   print "Off ";
   out AddrLpt% , 1
   delay .5
loop until instat 'Continue until any key is pressed
out AddrLpt% , 0  'Leave the port in a known state
```

operating in a command prompt window using DOS emulation, accesses to the printer port may fail since the Windows operating systems intercepts and blocks such calls. Windows does provide its own routines for accessing this port, but calling these requires programming in the more regimented Windows environment, and losing the ease of programming otherwise possible.

For a computer running a proper DOS operating system, the programme in Table 2.2, written in Power Basic, shows how to flash a LED connected (via a resistor) to pin 2 and ground. This listing also shows how to look up the actual address of the printer port, rather than assuming it follows the standard given above.

# The serial port

Most PCs come with one, or especially on older models two, serial or COM ports. Interfacing to these in software can be a lot easier than for the parallel port, but the serial format makes it slightly more difficult to make use of the data from this port. On most PCs the port takes the form of a nine-pin male D-type connector on the back panel, although on very early machines a 25-pin male connector was used. Four inputs and three outputs are present so, including ground, all pins are fully utilised on the nine-pin connector. For the transmission of data, only two of these are really of major importance; though others are used for handshaking and control, and can usually be hardwired or ignored by software if not needed for their designated purposes. The two are Transmit Data (abbreviated to 'TXD') and Receive Data (RXD). Data is transmitted one byte at a time (usually eight bits) serially by toggling the TXD line. The speed of data transmission is determined by how much time is allocated to each bit, and a wide range of standard rates is possible – usually called the *baud rates*. Values frequently range from 75, 150, 300 baud to 115200 baud, doubling each time with a few extra values thrown in. The exact signalling protocol used is ideally as defined in the *RS232* or, as it is now called, the *EIA RS232* standard. This is a complex

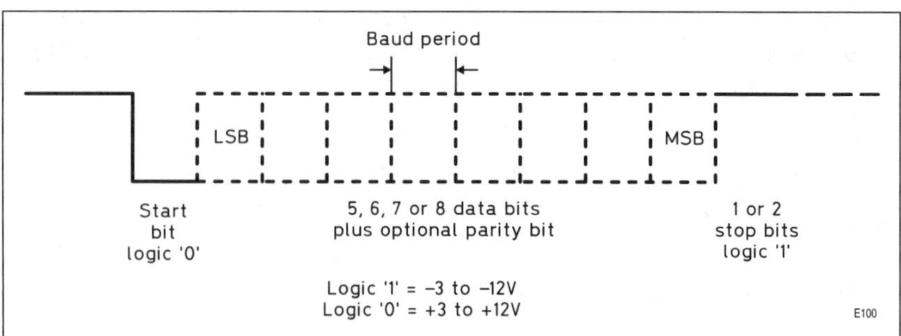


protocol document defined even before the days of PCs, and describes exactly how the data flow between the peripheral devices is controlled by the handshake lines, and the format of the data. The full protocol and most of the handshaking functions can usually be ignored, resulting in a much simplified connection, but whatever route is chosen the data always has the following form.

Each byte is preceded by a start bit, a logic '0', taking up one baud symbol duration. This is followed by either five, six, seven or eight data bits, sent least-significant bit first. A parity bit is sometimes added for error checking and the block of data is terminated with either one or two logic '1' stop bits as illustrated in Fig 2.1.

The huge variety of data formats alone makes the serial RS232 protocol a difficult 'universal standard'. For computer use, however, one or two variants have become more widely adopted. As we nearly always work with eight-bit bytes, one byte per frame is an obvious choice, and more often than not we are usually concerned with highly reliable communication between two items of hardware over wires, so parity checking is not needed. For maximum speed one stop bit is usually considered adequate. The baud rates are less standardised, but as older PCs could only cope with speeds equal to or less than 19200 baud, this rate is often used. The format is usually abbreviated to a form such as '19200 N81' which means 19200 baud, no parity, eight bits per character, one stop bit. Lower rates are often used when a radio or modem path is included in the link and speed is limited by the bandwidth of the link; conversely rates can go up to 115200 baud when trying to get the maximum amount of data possible into a PC. The maximum amount of data that can be transported for a given speed can be calculated as follows:

Characters/second = Baud rate / (1 start bit + Number of data bits + Parity + Number of stop bits)

So for the example given above, characters can be transmitted at the absolute maximum speed of:

19200/(1 + 8 + 2) = approximately 1745 bytes/second

One advantage of this asynchronous serial format is that as characters contain their own framing information, preceded by a start bit and terminated by at least one stop bit, they do not have to be sent continuously – the

**Table 2.3. RS232 pin connections and signals**

| Pin no 25-pin | 9-pin | Signal name | Direction of signal flow |
|---|---|---|---|
| 2 | 3 | Transmit Data (TXD) | DTE > DCE |
| 3 | 2 | Receive Data (RXD) | DTE < DCE |
| 4 | 7 | Request To Send (RTS) | DTE > DCE |
| 5 | 8 | Clear To Send (CTS) | DTE < DCE |
| 6 | 6 | Data Set Ready (DSR) | DTE < DCE |
| 7 | 5 | Ground (GND) | |
| 20 | 4 | Data Terminal Ready (DTR) | DTE > DCE |

hardware continuously searches for a start bit and only reads the character when the transition from logic '1' to '0' is detected. It is also possible for receiving software to determine the baud rate automatically. By storing the entire contents of the received signal waveform, software can search for transitions and measure the minimum distance between them. As the position and polarities of the start and stop bits are known, after a few characters have been received the correct baud rate can be deduced from timing measurements. This procedure is speeded up if a sequence of known characters is always present. An example of auto baud determination is seen when an external telephone modem is connected to a PC. Modems are controlled by commands sent over the RS232 interface that always begin with the letters 'AT'. The first time a modem encounters any signal transitions on the RS232 interface it knows the first two characters must be 'AT' and, by examining the pattern of '1's and '0's, can determine the speed of the interface.

## RS232 connections

RS232 was originally defined as a standard to interconnect a modem – the Data Communications Equipment (DCE) – to a Data Terminal Equipment (DTE), and pin allocation and labelling was made accordingly. The DCE has a *female* connector, either 25-pin or nine-pin, and the DTE has a *male* connector; leads for connecting a DCE to a DTE are connected in parallel, pin 1 to pin 1 etc, assuming the same connector type. Terminology for the Received Data (RXD) and Transmit Data (TXD) signals are as if they are used over a modem interface, ie a DTE puts out Transmit Data, and receives RXD. At a DCE port, this terminology can be confusing as TXD now reads in the data (from the DTE) and RXD sends it out! We need to be very aware of this potential for confusion, as peripherals designed to interface to a PC via the serial port are often configured as a DCE to enable a one-for-one connection.

The meanings of the signals and associated pin connections are given in Table 2.3. Certain other connections are also available on some 25-pin interfaces that are not used with the later nine-pin version.

By adhering to the DTE/DCE convention, connections, connector polarity, signal names and data flow is logical and (almost) intuitive. Things get rather more complicated when two DTE-type equipments need to talk

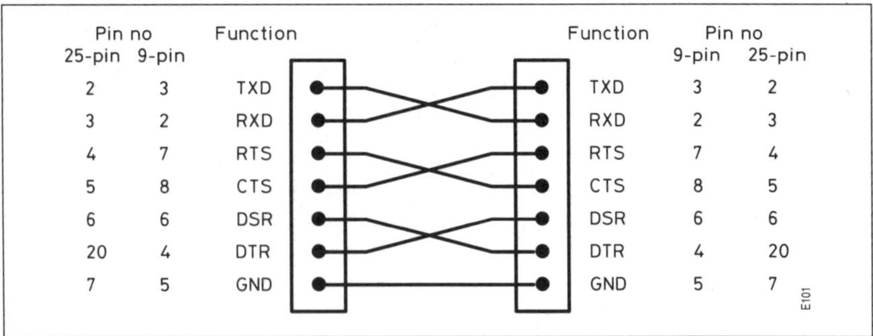

Fig 2.2. Pin connections for an RS232 null-modem cable

to each other. An example of this is when trying to connect two computers via their serial ports – for example, to swap files between machines with completely different operating systems. Here a connection type known as a *null modem* is needed – so called because it behaves as if replaces a pair of modems. A null modem cable has a female connector at each end and, as a minimum, has the TXD and RXD connections swapped over plus ground. Usually RTS and CTS are crossed over, and frequently DTR and DSR. This way, each DTE can now provide the handshaking signals required for a full intercommunication protocol. More often, however, the full handshake is not provided, and CTS is often just connected to RTS, and DTR to DSR locally at each end, so each computer in effect provides its own handshake. See Fig 2.2.

Further complications are sometimes added when manufacturers of some equipment provide RS232 interfaces with the wrong polarity of connector. In one case of recent experience, considerable time was wasted finding cable connections to interface to a professional HF receiver that purported to have a 25-way RS232 connection on the back. This was configured as a DCE (sensible) *but* had a male connector, required RTS/CTS handshaking, and did not provide a suitable signal to activate the controlling PCs DSR input. The moral is, to paraphrase from reference [1], "Once you have found an RS232 lead that works for your particular application, guard it with your life."

### Voltage levels for RS232 interface

RS232 was originally defined as follows: logic '1' equals –3 to –12V, logic '0' equals +3 to +12V. The region from –3 to +3V was undefined and constituted an error. Therefore the 'resting' state of an RS232 interface, when active but not sending data, has a negative voltage on the data line, corresponding to the state of the last stop bit (a logic '1'), and usually a positive voltage on the DTR line for a logic '0', indicating the serial port is active.

The presence of both positive and negative voltages on various pins when the port is in use can lead to some interesting short cuts when using it for interfacing. Although the threshold for detecting the presence of a '0' or '1' would ideally be at zero volts for optimum discrimination of the two voltage levels as defined above in the presence of noise, it is usually set at around +1.5V in most RS232 receiver chips. The reasons are twofold. First (and probably historically), it allows an open-circuit line to be treated as if it were negative, ie at logic '0', which in the case of the TXD/RXD lines is an error condition, and in the case of any of the handshaking lines represents an inactive state – so an open-circuit condition cannot give a false indication. The second reason is that in the early days of home computers – the BBC Computer era – manufacturers economised on the serial interface by using just the logic levels of 0/5V rather than having to introduce a negative supply and higher voltages. Having the 'true' RS232 receiver chips switch at +1.5V meant that this much simpler interface, which could be as simple as a single TTL or CMOS buffer as the line driver, would interface with peripheral devices that employed the 'true' signalling voltage levels.

### Driving the serial port

Unlike the parallel port, it is usually not necessary to directly access the PC hardware in order to gain control of the COM port(s). All programming languages have commands or routines within them to open these ports, set the operating parameters and allow data to be written to or read from them. This allows the serial interface to be used with any operating system from DOS, any variant of Windows or any OS of choice. Table 2.4 is a listing in Power Basic showing how the COM port can be used to receive serial data and print it to the screen. Table 2.5 illustrates how data may be sent over this interface. Some programming languages even include hardware handshaking as part of the serial port setting-up commands. If direct access to the hardware is possible, the serial port can be used in a number of non-standard ways, for example using the RTS line as a CW keyer output to drive a transmitter.

If it appears that undue attention is being paid to a slow old-fashioned interface, remember that RS232 is the only 'standard' that has truly endured across every computer since the dark ages, and is probably the only standard that can be guaranteed to be compatible across computer platforms for a long time to come. Its serial nature allows a natural connection for radio-based applications, and it is simple to design interface hardware, as numerous projects later in this book will show.

# The ISA bus

Until the very latest machines came along, all PC motherboards had at least one spare PCB socket that gave direct access to the PC's data, address and control lines. This means that it is possible, with suitable address decoding and interfacing circuitry, to add hardware to a PC that allows data to be directly read from, or sent to the processor, giving the fastest possible transfer mechanism. The ISA (Industry Standard Architecture) bus is recognised by its double-sided 31-way PCB-type connector next to a smaller 18-way connector as shown in Fig 2.3.

The larger one can be used alone, and gives access to just the lowest eight data bits with address lines A0 to A19 and all the control signals needed – it

**Table 2.4. Programme to receive data over the serial link and print to the screen, with identification of the I/O port address**

```
print "Example programme to receive data over the serial link"
          'Prints data directly to screen


                          'First of all, find hardware location of COM1
def seg = &h40            'Data table in BIOS memory
AddrCom% = peeki(0)       'Entry in table for COM1 port I/O address
print hex$(AddrCom%)      'Just show the address as we don't actually need it here


open "COM1:9600,N,8,1,DS,CS" as 1
                          'Open COM 1 serial port as file reference 1
                          '9600 baud, No parity, 8 data bits, 1 Stop bit,
                          'Ignore DSR and CTS handshaking lines (for simple interfacing)

on com(1) gosub GetSer    'Generate interrupt each time a character is received
com(1) on                 'Turn on the interrupt


do                        'Somewhere to idle away the hours while
loop until instat         ' waiting for an interrupt.
                          ' Just loop until any key is pressed
end

GetSer:                   'Serial port interrupt handler
   l% = loc(1)            'Look to see how many characters have arrived
   c$ = input$(l% , 1)    'Get all new characters and read into string c$
   print c$;              'Print all received data to screen
return
```

**Table 2.5. Programme to send data over the serial link**

```
print "Example programme to send data over a serial link"
                  'Echos any character typed on the keyboard to COM 1

open "COM1:9600,N,8,1,DS,CS" as 1
                  'Open COM 1 serial port as file reference 1
                  '9600 baud, No parity, 8 data bits, 1 Stop bit,
                  'Ignore DSR and CTS handshaking lines (for simple interfacing)


do
   while not instat : wend    'Wait for any keyboard key to be pressed
   c$ = inkey$                'Get the character just entered
   put 1 ,, c$                'Send it to the serial port
loop until c$ = chr$(27)      'Keep doing this until [esc] key is pressed

end
```

derives from the very first PCs that had just an eight-bit data bus and 1 megabyte of memory space. The smaller connector gives access to the upper byte of the data bus, allowing full 16-bit access as well as a few more control signals; 32-bit access is not possible using the ISA bus. Table 2.6 gives a description of the signals present on this bus.

Connecting via the ISA bus makes it feasible to build hardware that uses spare I/O space for independent input and output, or alternatively to use spare memory space for faster memory-mapped I/O or direct memory access (DMA). These latter techniques are too complex to cover here but interfacing details can be found in reference [2].

Fig 2.4 shows the circuit diagram of a simple eight-bit latched input/output stage which is mapped to a spare I/O space at address 0x300 – the I/O space from 0x300 to 0x31F is allocated for user hardware.

Data is written to this user interface simply by writing a byte to I/O address 0x300 where it is latched into the output flip-flop. Data may be read from the input lines by reading from this I/O address.

Later PCs have a smaller edge connector, giving access to the full 32-bit bus (this is the modern PCI interface) but using this interface is not nearly so straightforward as the ISA bus. The PCI interface has been defined to make use of the plug-and-play capability of operating systems to automatically recognise new hardware.

Connecting to the PCI bus is beyond the scope of this book and requires custom chips, but full details are available in reference [3].

## Timing issues

One area that the basic PC is not good at is running real-time software! While reasonably accurate date and time keeping is available using the internal real-time clock, generating intervals of micro or milliseconds is near-impossible with any accuracy. 'Raw' DOS includes a software timer that generates an interrupt every 55ms approximately and is part of the operating system, and timing can generally be defined to an accuracy of this order. In fact, several programming languages include

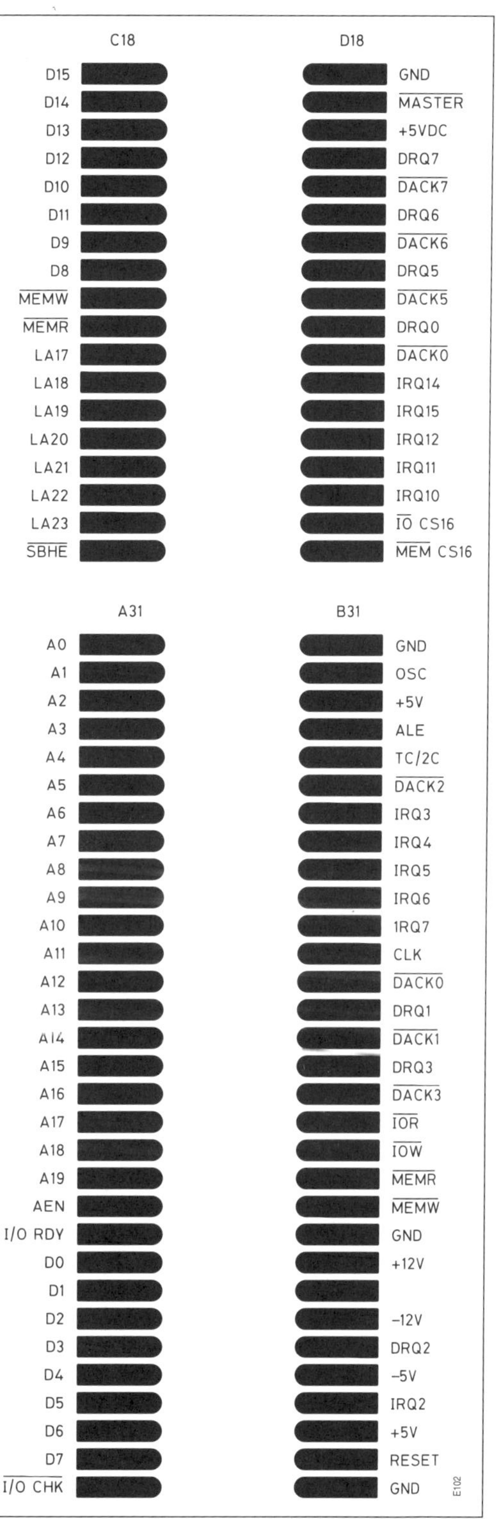

*Right:* **Fig 2.3. ISA bus pin-out**

| Table 2.6. ISA bus (Industry Standard Architecture) signal descriptions | |
| --- | --- |
| SA0 to SA19 | *System Address* bits 0 to 19 are used to address memory and I/O devices. Only the lower 16 bits are used during I/O operations to address up to 64k I/O locations. SA0 is the least significant bit. SA19 is the most significant bit. These signals are gated on the system bus when BALE is high and are latched on the falling edge of BALE. They remain valid throughout a read or write command. These signals are normally driven by the system microprocessor or DMA controller but may also be driven by a bus master on an ISA board that takes ownership of the bus. |
| ALE | *Address Latch Enable* (also sometimes called BALE = Bus Address Latch Enable). The address bus is latched on the rising edge of this signal. The address on the bus is valid from the falling edge of ALE to the end of the bus cycle. |
| AEN | *Address Enable* is used to degate the system microprocessor and other devices from the bus during DMA transfers. When this signal is active the system DMA controller has control of the address, data, and read/write signals. This signal should be included as part of ISA board select decodes to prevent incorrect board selects during DMA cycles. |
| SD0 to SD15 | *System Data* serves as the data bus bits for devices on the ISA bus. SD15 is the most significant bit. SD0 is the least significant bits. SD0 to SD7 are used for transfer of data with eight-bit devices (eg this lab). SD0 to SD15 are used for transfer of data with 16-bit devices. |
| −IOR | *I/O Read* is driven by the owner of the bus and instructs the selected I/O device to drive read data onto the data bus. |
| −IOW | *I/O Write* is driven by the owner of the bus and instructs the selected I/O device to capture the write data on the data bus. |

functions that generate interrupts at multiples of this interval such as the ON TIMER command in all dialects of Basic. Measurement of events such as I/O lines changing is sometimes possible to microsecond accuracy using a microtimer, but this in turn falls off in accuracy as time intervals approach 55ms.

The Windows operating system, including the command prompt window, is much worse still. As Windows is a multitasking operating system, it is continuously servicing all the independent pieces of software running, and so no guarantee can be provided that machine resources will be available to the wanted software at the required intervals. By operating a command prompt window at full screen, most of the PCs resources can be directed to the software and a reasonable approximation to real-time operation is usually possible – subject to the 55ms timing issues.

The 55ms timing interval derives from dividing an hour by 65536. Quite why the timing interval should be derived like this is by now lost in the folklore of computer history, but for accurate timing purposes the exact interval is given by $t = 3600/65536$ seconds, or 54.93164ms. This of course is subject to the accuracy of the oscillator used for the PC's timing clock – this is often just a packaged crystal oscillator and easily subject to an error of 50 parts-per-million or worse.

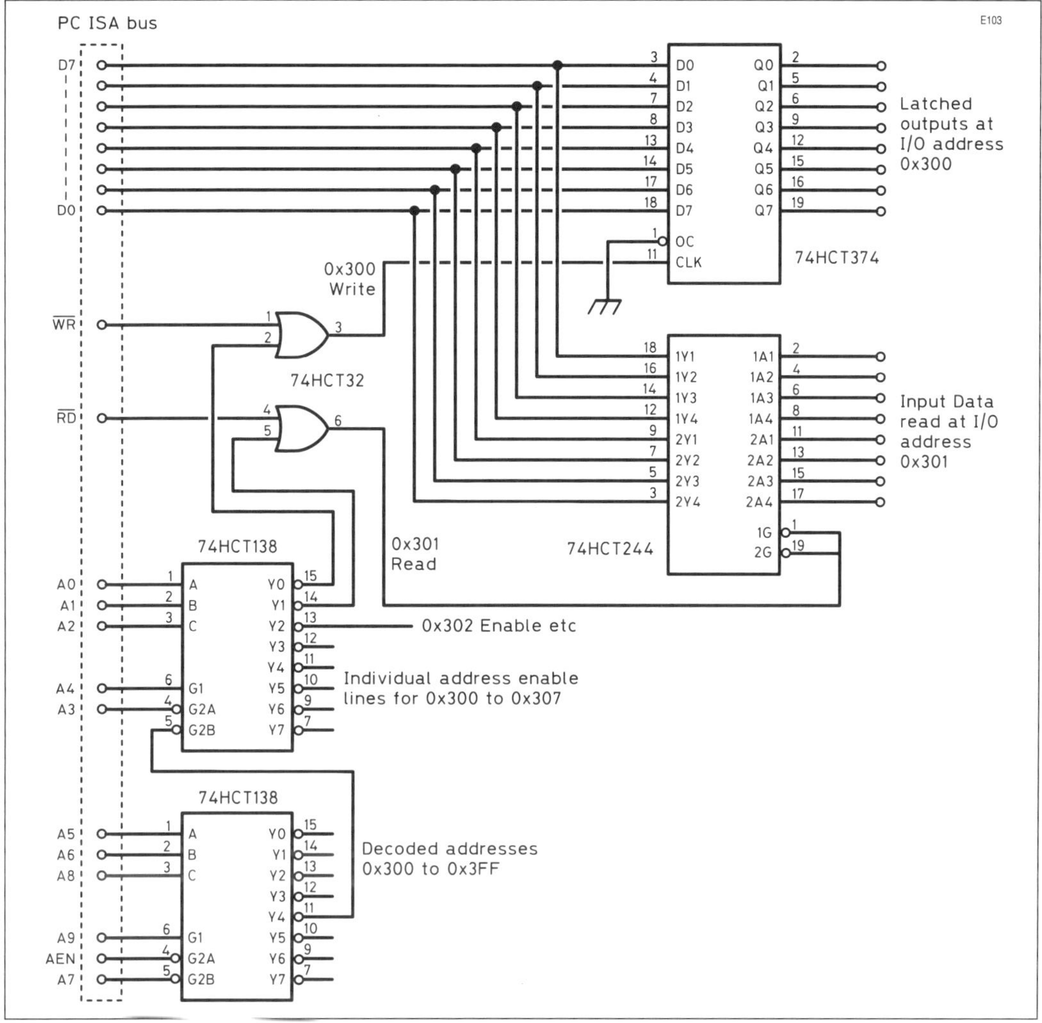


Fig 2.4. Circuit diagram of a eight-bit I/O latch on ISA bus

## CW keyer programme

The Power Basic listing given in Table 2.7 is for a CW keyer for a beacon transmitter that sends a user-entered message at predetermined intervals. The message, CW speed and repeat interval can be changed arbitrarily while the programme is running.

The transmitter is keyed via the DTR line of the RS232 interface and the RTS line is used for transmit/receive control if this facility is wanted. As keying speeds (CW dot lengths) that are exact multiples of 55ms do not give a useful range of CW speeds, the programme shows one way of generating reasonably accurate time delays that are not restricted to multiples of this interval. A software delay is generated in a loop and at the start of the programme this delay loop is calibrated by making use of the microtimer instruction. The calibration constant so determined is then used to generate an accurate 1ms time delay. This technique is quite good when used on

**33**

**Table 2.7. CW Beacon, or test keyer, programme. Sends CW equivalent of a message typed in, at user specified speed and repeat interval**

```
cls
dim charray$(127)
calibrate = 1
interval  = 1                    'needed initially to calibrate delay loop

mtimer                            'Initialise the microtimer            .
call pause(500)
tim = mtimer                      'mtimer command measures no. of us for 500 loops
calibrate = 500000 / tim        'number of original loops for 1 ms.
          'Calibrate constant now defined

'Now define CW characters
data  01,1000,1010,100,0,0010,110,0000,00,0111,101,0100,11,10,111,"0110"
data 1101,010,000,1,001,0001,011,1001,1011,1100                        'A - Z
data  "11111","01111","00111","00011","00001","00000","10000","11000","11100","11110"'0 - 9
'         Special case for punctuation, stored with character
data  ".","010101","/","10010","?","001100",",","110011","=","10001","-","100001"


for a% = 65 to 90          'ASCII A-Z
    read charray$(a%)
    charray$(a% + 32) = charray$(a%)        'lower case
next a%

for a% = 48 to 57          'ASCII 0-9
    read charray$(a%)
next a%

for z = 1 to 6                        'allowed punctuation symbols
    read sy$,charray$(asc(sy$))
next z

if dir$ ("KEYER.INF") > "" then '.INF file saves COM port in use
    open "i" , 1 , "KEYER.INF"
    input #1 , cport%
    close 1
else
    input "COM Port for interface  ",cport%
end if

if cport% = 0 then cport% = 1
if cport% > 4 then cport% = 4    'COM Port > 4 not allowed
def seg = &h40     'CORRECT way to find location in I/O space, address table in segment &h40
mcraddr% = peeki(cport% * 2 - 2)              'Location of UART base I/O address
if mcraddr% = 0 then print "No COM";cport%;"port at this address" : end

mcraddr% = mcraddr% +  4  'now point at modem control register
out mcraddr% , 0
print "Transmit test keyer - stays in Key Down state during pause"

keyerbit% = 1    'RTS Keys transmitter for compatibility with VE2IQ 'COHERENT'
txbit% = 0       'DTR Controls Tx/Rx switching
locate 3,1
```

**Table 2.7** *(continued)*

```
'Keyer uses serial port COM";cport%;"for interface
'Output from RTS and DTR lines,   Polarity: +ve = key down / Tx active
'25 Way connectors    -    pins 4 and 20      Gnd pin 7
'9 Way connectors     -    pins 7 and 4       Gnd pin 5

locate 4 , 10
input; "Speed  wpm            ",speed
if speed = 0 then speed = 18 : print "18"
oldspeed = speed

locate 5 , 10
Input; "Repeat delay         ",repeatime
oldrepeat = repeat

locate 9 , 10
print tab(10);"Allowed punctuation    ";
color 15
for a% = 33 to 127
     if (a% < 48 or (a% > 57 and a% < 65) or (a% > 90 and a% < 97) or a% > 122) and
charray$(a%) > "" then
            print chr$(a%) + " ";
     end if
next a%
color 7

locate 11 , 10
line input "Text   ",tx$
tx$ = ucase$(tx$)
if tx$ = "" then tx$ = "TESTING TESTING 123"
locate 11 , 16
print tx$
'User prompt to show keyboard characters that allow change of speed,
'repeat interval and text while  programme is running

locate 25 , 15
color 12 , 1
print "    [esc]";
color  7:print"    ";
color 12:print "S";
color  7:print "peed      ";
color 12:print "R";
color  7:print "epeat interval   ";
color 12:print "T";
color 7 :print "ext     ";
color 7 , 0

do
     interval = 1200 / speed      'dot clock interval ms
     incr rno&
     locate 15 , 30
     color 10
     bit set mcrdat% , txbit%
     out mcraddr% , mcrdat%
```

**Table 2.7 *(continued)***

```
    print "Sending";
    locate 16 , 1
    for t% = 1 to len(tx$)
          curchar% = asc(mid$(tx$ , t% , 1))
          if curchar% = 13 or curchar% = 32 or curchar% = 8 then
                  call pause(2)
                  print " ";
          else
                  pattern$ = charray$(curchar%)
                  if pattern$ = "" then iterate for
                  for symbcount% = 1 to len(pattern$)
                          dotdash$ = mid$(pattern$ ,  symbcount% , 1)
                          bit set mcrdat% , keyerbit%
                          out mcraddr% , mcrdat%
                          if dotdash$ = "0" then
                                  print ".";
                                  call pause(1) 'send dot
                          else
                                  print "-";
                                  call pause(3) 'send dash
                          end if
                          bit reset mcrdat% , keyerbit%
                          out mcraddr% , mcrdat%
                          call pause(1)          'Intersymbol gap
                  next symbcount%
                  call pause(2)                 'letter gap
                  print " ";
          end if
    next t%
    if instat then ik$ = ucase$(inkey$) 'Check for keyboard hit & read char.
    locate 16 , 1
    print space$(80);
    locate 17 , 1
    print space$(80);

    if ik$ > "" then           'Take action on any user intervention
          locate 15 , 30
          print space$(20)
          select case ik$
          case "R"              'Request a new repeat interval
                  color 7
                  locate 5 , 10
                  print space$(60)
                  locate 5 , 10
                  Input; "Repeat interval  (s) ",repeatime
                  if repeatime = 0 then
                          repeatime = oldrepeat
                          locate csrlin , pos - 1
                          print repeatime
                  else
                          print
                  end if
                  oldrepeat = repeatime
                  ik$ = ""
```

**Table 2.7** *(continued)*

```
            case "S"              'Request a new speed
                 color 7
                 locate 4 , 10
                 print space$(60)
                 locate 4 , 10
                 Input; "Speed  wpm            ",speed
                 if speed = 0 then
                         speed = oldspeed
                         locate csrlin , pos - 1
                         print speed
                 else
                         print
                 end if
                 oldspeed = speed
                 ik$ = ""
            case "T"              'Request new text to send
                 color 7
                 locate 11 , 10
                 print space$(60)
                 locate 11 , 10
                 line input "Text  ",tx$
                 tx$ = ucase$(tx$)
                 if tx$ = "" then end
                 locate 11 , 16
                 print tx$
                 ik$ = ""
            end select

    else            'Otherwise sit in loop looking for next repeat interval
            locate 15 , 30
            color 12
            print "Waiting"
            bit reset mcrdat% , txbit%
            bit reset mcrdat% , keyerbit%
            out mcraddr% , mcrdat%
            call pause(3)
            out mcraddr% , mcrdat%
            delay repeatime
            bit set mcrdat% , txbit%
            out mcraddr% , mcrdat%
            call pause(3)
    end if
    if instat then ik$ = ucase$(inkey$)
loop until ik$ = chr$(27)        'Escape key to get out
out mcradr% , 0
end

sub pause(ms%)
    shared calibrate , interval                 'Global variables
    for a& = 0 to calibrate * ms% * interval    'This loop generates ms% * 1ms
        'delay once calibrated
        next a&
end sub
```

**Table 2.8. Connections for the game port, 15-way D-type female connector**

| Pin | Function |
| --- | --- |
| 1 | Potentiometer common Joystick A (+5V) |
| 2 | Button 1, Joystick A |
| 3 | X Co-ordinate potentiometer for Joystick A |
| 4 | Button common Joystick A |
| 5 | Button common Joystick B |
| 6 | Y Co-ordinate potentiometer for Joystick A |
| 7 | Button 2, Joystick A |
| 8 | Unused |
| 9 | Potentiometer common Joystick B (+5V) |
| 10 | Button 1, Joystick B |
| 11 | X co-ordinate potentiometer for Joystick B |
| 12 | Used for MIDI interface only, Transmit |
| 13 | Y co-ordinate potentiometer for Joystick B |
| 14 | Button 2, Joystick A |
| 15 | Used for MIDI interface only, receive |

older, slower machines running true DOS, and can give quite acceptable results in a full-screen command prompt window.

# Other ports

There are a few other ports on a PC that are accessible for user input/output.

### Game port

One of the oldest is the game port, or joystick port – this is usually accessible via a 15-way D-type connector and, as its name implies, is intended for the connection of two joysticks.

Each joystick contains two potentiometers, making a total of four that can be connected, plus two press-button switches. Each potentiometer controls the delay on one of four monostables, and the delay can be read by software, as can the status of the press buttons. The game port is probably of little value for any other functions than what it was designed for, but the press-button lines could be used for single-bit status monitoring of some experimental equipment without having to tie up either the COM or parallel ports.

One possibility would be for monitoring the squelch status of an FM receiver while scanning a band for activity – see the next chapter. Connections for the game port are shown in Table 2.8. The buttons can be read directly by reading port 0x201 and looking at bits 4–7; no triggering is required.

### Universal Serial Bus

The USB interface is one of the latest connections to appear on modern PCs. It is only accessible through Windows, and requires users to make use of libraries and operating system calls for the PC driver software. However, USB is extremely versatile and capable of speeds of several megabits per second. At the peripheral end, a range of dedicated USB chips now exist that make interfacing to this port straightforward. Also, USB driver 'cores' are available within some microcontroller chips, the intention being to make operation using USB as transparent to hardware designers as possible. Full details of the USB specification are available from reference [4]. Suitable interfaces are given in references [5] and [6]. USB peripherals can be purchased that add extra parallel or serial ports – although when installed, these are only usually available for Windows software and cannot be used through a command prompt window. By using the USB interface for the system printer, the LPT port can be freed up for interfacing as described earlier.

### The sound card

Apart from its obvious use for input and output of audio signals, the sound card can be treated as an analogue I/O port. Its frequency response does not extend down to DC, rolling off at a few tens of hertz usually – so it is really only suitable for audio-like signals. However, at its fastest sampling rate of 44kHz it can cope with signals up to 20kHz, making it suitable for looking at portions of the RF spectrum that could contain many signals. Refer to Chapter 10 for use of the sound card.

# Isolation and EMC issues

Electrically, all computers can be considered as 'dirty' items of equipment. All generate spurious signals that are multiples of their many internal oscillators, rapidly switched data and address lines, and harmonics from switched-mode power supplies. Here, modern machines are a lot better than their older counterparts as the modern EMC legislation they now have to meet is quite strict.

However, much can be done by users to minimise coupled interference. In most cases the PC itself is quite well screened, although some branded makes are very much better than others here. Most interference problems come from the very interconnections we need to make our interfaces. RF signals generated within the case are coupled onto the I/O ports and so to the outside world.

The use of screened cable for connections to the serial and parallel ports can do a lot to minimise radiation of these coupled signals. The screen must be firmly bonded to the chassis at each end to be effective but, when these cables connect to sensitive radio receivers, for example, it can still be difficult to keep the spurious signals down to manageable levels. Bypass capacitors and ferrite beads strategically placed can do a lot to help, provided they do not degrade the interface signal waveform.

Depending on the application, there are a few interface circuits that can be employed to drastically cut down coupled signals. Optical coupling is a very useful trick to both remove ground loops and common-mode RF

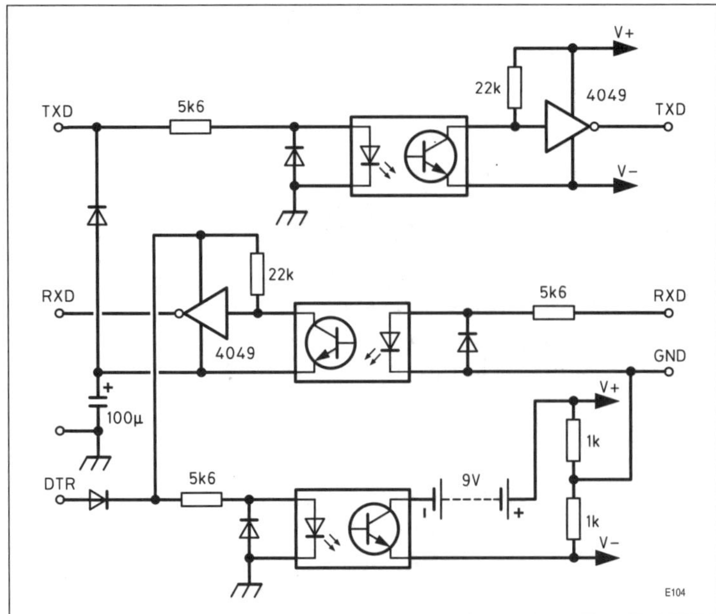



**Fig 2.5. Opto-coupled RS232 interface**

signals. Fig 2.5 shows a simple opto-coupled interface for the RS232 data lines. A separate battery is needed to supply power for the PC side of the interface but this can be switched automatically from the PC end via DTR, making use of another opto-isolator, as shown. Sensitive relays that can be directly driven from the power available on the serial port can also be used.

Small 1:1 isolating transformers can prove useful for audio interfacing to the sound card. These are usually designed for $600\Omega$ systems, but work adequately for this task. More details on interfacing the sound card are given in reference [7].

# References


[1] *RS232 Made Easy*, Martin D Seyer, PTR Prentice-Hall Inc, 1991. This book provides a firm background in data communications environments and gives the reader the ability to connect printers, terminals and computers using the RS232 interface. It discusses communication terminology, details RS232 operation in dial-up and private-line environments, and the use of null modems. It equates complex technical aspects of computers, communications equipment and peripherals to an easily understood railway system. This 436-page book eliminates the tedious research usually needed to properly connect two devices. It provides charts which give pin assignments for over 200 devices, with connection diagrams, explanations of option settings, and excerpts from the Electronics Industrial Association Standard.

[2] More details on the ISA bus can be found at http://wearcam.org/ece385/lecture6/isa.htm#1.0.

[3] PCI Bus Specifications is available on request from the PCI website: www.pcisig.com.

[4] The USB Specification is obtainable by visiting www.usb.org.

[5] Future Technology Devices International (FTDI) manufactures a large range of USB interface chips. See their website at www.ftdichip.com/.

[6] '16 bit I/O via USB', *Electronics World* April 2002, pp12–18.

[7] *Digital Modes for all Occasions*, Murray Greenman, ZL1BPU, RSGB, Chapter 7.

# CHAPTER 3

# Transceiver control and band monitoring

**M**ost modern amateur radio receivers and transceivers have the facility for remote control of all the front-panel functions. By connecting a transceiver to a PC or other type of controller, a whole new world can be opened up of automatic band tuning and scanning, automatic frequency selection, beacon monitoring and spectrum analysis, to name but a few.

## Transceiver interfaces

Almost all modern transceivers have an interface for computer control, and there is a wide variety of software available to communicate with the microprocessor inside the rig that controls the frequency synthesiser and many other functions. The following section is taken from reference [1].

The PC software exchanges data with the transceiver using one of the PC's serial interface or COM ports. The COM port uses the RS232 interface for exchanging serial data with the outside world. The problem is that most transceiver serial ports provide TTL-level signals: +5V for logic '1' and 0V for logic '0'. Fig 3.1 shows why an interface is needed in order to shift these voltage levels between the two systems. Controlling a transceiver by a PC is usually a two-way data exchange, so each direction needs its own level shifter.

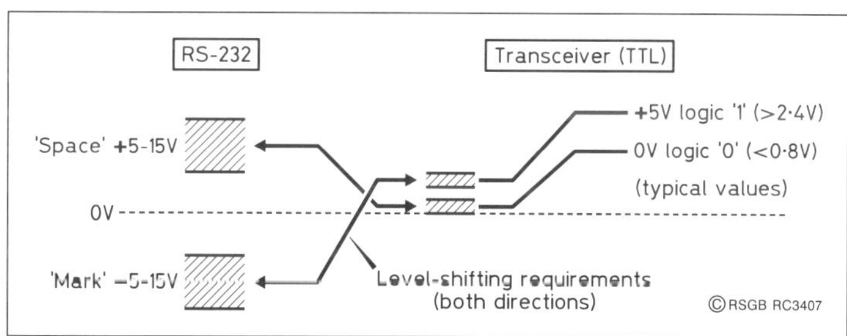


Fig 3.1. Level-shifting requirements between RS232 and TTL

A few transceivers now provide direct plug-and-play RS232 compatibility, using a nine-pin D-type connector which is similar to the PC's COM port. All others need an interface, so let's examine the connector pins at the COM port to see what's needed. The DCE/DTE conventions as defined in Chapter 2 do not apply here, and connections are not necessarily via a null modem cable either. The 'Connects to' column shows how all the lines from the PC connect to a different-named line at the other end of the link,

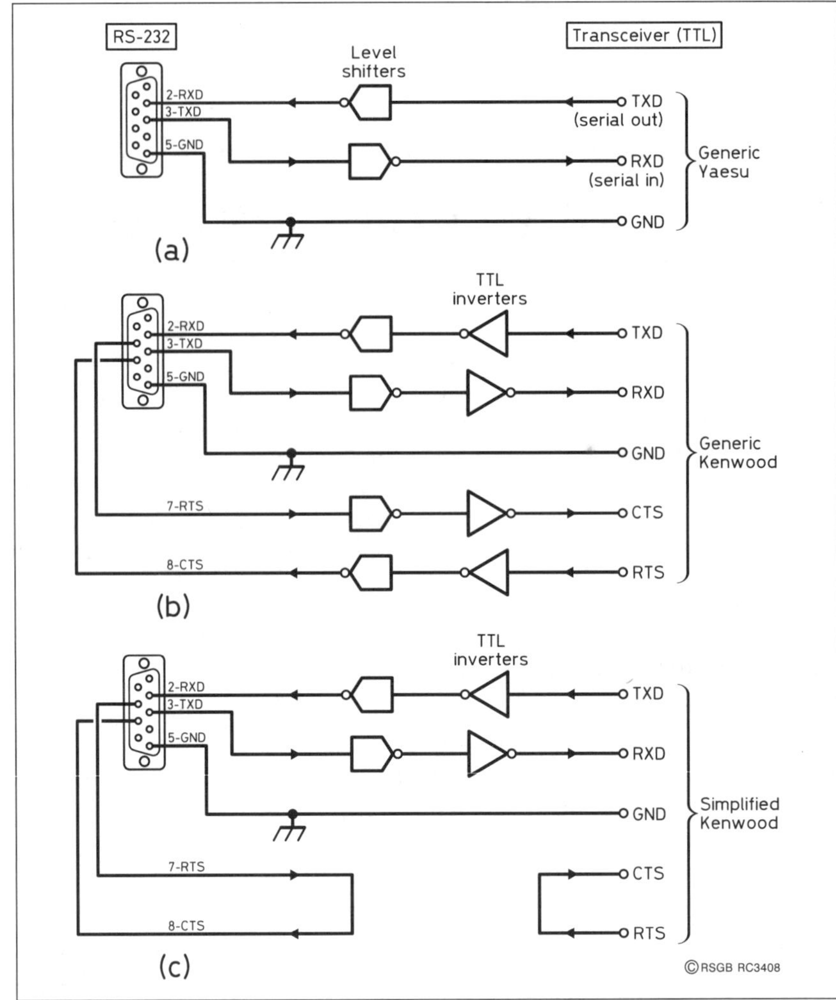



with the single obvious exception of ground. The most important pair is TXD and RXD. The TXD output from the PC carries transmitted data from the PC to the transceiver, and is connected via the interface to the transceiver's RXD input (sometimes called 'Serial In'). The transceiver's TXD output (or Serial Out) sends data back to the PC's RXD port. The 'Big Three' transceiver manufacturers each have different control protocols and different hardware interface requirements, and these also vary between different transceivers from the same manufacturer. The control protocols are usually handled for you by the software authors – and now, here's how to handle the hardware requirements.

All that's absolutely needed for two-way serial communication are just three wires: TXD, RXD and ground. That's what Yaesu rigs use (Fig 3.2(a)). In this kind of 'free streaming' communication, either end can send data whenever it wants, and this works fine for the kinds of simple, direct links we need for transceiver control. The Kenwood interface is more complex:

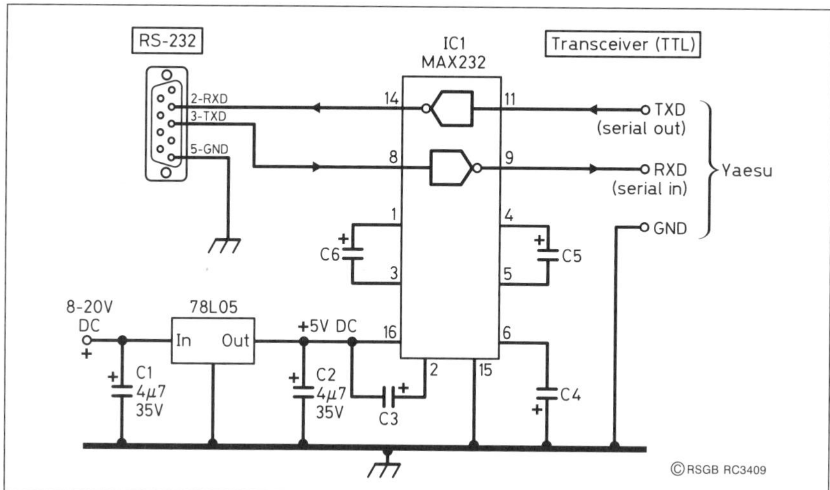

Fig 3.3. Practical Yaesu interface using MAX232. The values of C3–C6 are dependent on the variety of MAX232 chip used (see the data sheets)

it uses reversed TTL logic polarities (indicated by the TTL inverters in Fig 3.2(b)) and it also requires Request To Send (RTS) and Clear To Send (CTS) handshaking on two additional lines. This RS232 jargon means that no data is supposed to be transmitted until the other end has been asked if it's ready to receive (RTS), and has sent back a confirmation (CTS). Fortunately RS232 communication doesn't always insist on such formalities. The hardware handshaking needs of Kenwood rigs can usually be satisfied by linking the RTS output directly back to the CTS input on the same connector (Fig 3.2(c)) – in effect, each device tells itself to go ahead.

Next let's look at some hardware for level shifting. Remember that the aim is to translate the TTL levels from the transceiver into RS232 (or compatible) levels for the PC (Fig 3.1). The first practical interface (Fig 3.3) uses a MAX232 IC to generate true RS232 voltage levels. To fall securely inside the RS232 specification, this requires DC supply rails at about +10V and −10V. The MAX232 revolutionised TTL/RS232 interfacing by generating both of these voltages on-chip from a standard +5V TTL supply, which makes the whole design very simple. The MAX232 contains two level shifters in each direction, so it can provide a more complete RS232 interface with handshaking if required. Fig 3.3 shows a MAX232 interface for Yaesu transceivers such as the FT-980 and FT-990 that don't already have a PC-compatible COM port. Wire it up in a small shielded box close to the transceiver, apply +12V (which the 78L05 regulates down to +5V) and away you go with real RS232 bipolar signalling. More elaborate interfaces for Yaesu, Icom and Kenwood rigs using the MAX232 are described in recent editions of the *ARRL Handbook*.

However, interfaces can often be much simpler than that. We don't really need true bipolar signalling, because the RS232 receiver ICs used in PC COM ports don't insist on the full-specification voltage levels. The noise margins are much reduced compared with true bipolar RS232 with its 20V swing, but they are generally adequate for short screened runs between a PC and a transceiver. Fig 3.4 shows an example for Yaesu rigs. R1 connects

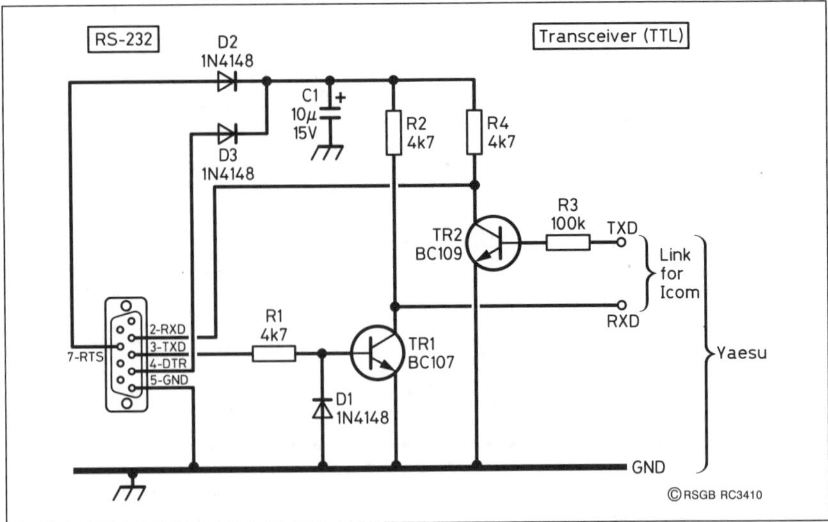

**Fig 3.4. Practical Yaesu/Icom interface deriving power from the COM port. Note the link for the Icom 'single wire' data bus**



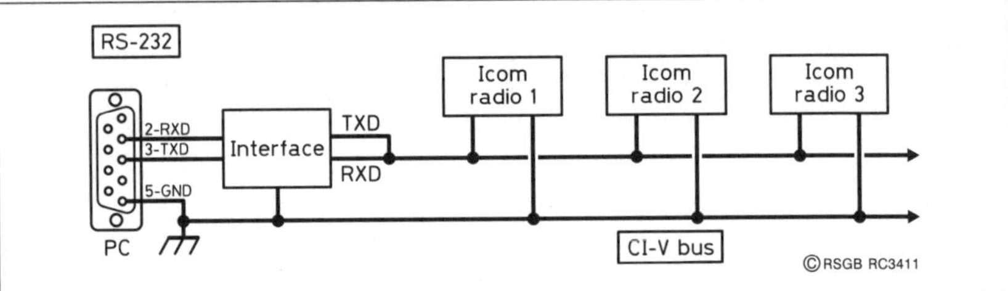

**Fig 3.5. Icom 'single wire' (plus ground) CI-V data bus allows multiple rig control; also used by Ten-Tec**



the PC's TXD port to the base of TR1, and a positive RS232 'space' level will pull TR1's collector voltage down to ground. The negative RS232 'mark' condition biases TR1 into cut-off (negative base voltage is limited by D1) so that R2 can pull the output line up to a valid TTL logic '1' level. TR2 works in a similar way: whenever the transceiver puts a TTL logic '1' on the line, TR2 pulls the RXD voltage at the PC COM port almost to zero, which the PC's RS232 receiver IC interprets as a valid 'mark'.

Icom rigs are different because they use a simple 'one wire' interface (actually a single wire plus ground screen). The Icom CI-V control protocol is quite sophisticated: it not only handles signalling both ways along the single wire, but also allows the PC interface and several Icom rigs to *share* that one data line (Fig 3.5). The CI-V system sends individually addressed data packets to each rig, and likewise recognises the identity of each rig when receiving data. The Icom hardware interface thus requires the input and output to be commoned – in Fig 3.4 simply link TXD and RXD at the transceiver side as shown. Examples of some of the Icom control codes for the IC746 are given in Table 3.1.

The two-transistor interface in Fig 3.4 needs no external power supply. The positive rail is generated directly from the RS232 port via D2 and D3, which are connected to the RTS and DTR lines. Either or both of these lines usually sits at the RS232 'space' level of about +10V. C1 stores this

| Table 3.1. Some of the Icom control codes for the IC746 transceiver |
| --- |

Controller to IC-746 (all codes in hexadecimal):

| FE | FE | 56 | E0 | Cn | Sc | Data area | FD |
| --- | --- | --- | --- | --- | --- | --- | --- |

IC746 to controller:

| FE | FE | E0 | 56 | Cn | Sc | Data area | FD |
| --- | --- | --- | --- | --- | --- | --- | --- |

Some example commands are shown below. For the complete command table and a more complete description, refer to the transceiver handbook.

| Cn | Sc | Description |
| --- | --- | --- |
| 00 | | Sets frequency (in data area) |
| 01 | xx | Sets mode (transceive) |
| 05 | | Sets frequency |
| 06 | 00 | LSB mode |
| 06 | 01 | USB mode |
| 06 | 03 | CW mode |
| 0F | 00 | Turns split operation OFF |
| 0F | 01 | Turns split operation ON |
| 0F | 10 | Selects simplex operation |
| 16 | 00 | Sets preamp OFF |
| 16 | 12 | Sets AGC ON, Slow |
| 1B | 00 | Sets the tone frequency for repeater use |

Numbers in the data area are sent in packed binary coded decimal form, two digits per byte. For example a frequency of 144.57500MHz would be coded as:

00  75  45  14

voltage and takes care of any minor gaps in the supply. Stealing a small positive and/or negative supply from the RS232 port is a well-known technique, but is limited to a few milliamps only, so it won't work with the basic MAX232 circuit in Fig 3.3.

An opto-isolated RS232 interface was shown in Fig 2.5. If you have multiple transceiver-PC interconnections including audio, these interfaces can be very useful because they also break the ground connection, which may help prevent hum loops.

## Watchdog timer for computer control of transmitters

Using a PC to control transmit/receive switching, such as is usual for most data mode software, could allow the transmitter to stay in transmit mode indefinitely if there was a software crash or malfunction. The only solution is then to reboot the computer, remove the interface connector or switch off the power to the rig.

Most data mode software interfaces to the PTT line by making use of the COM port. The circuit in Fig 3.6 is for a watchdog timer that during normal operation passes the DTR/RTS state straight though to the transistor controlling the PTT line. However, with the component values shown, after approximately one minute of transmitting the PTT line is released and the LED lights up to show an error or timeout condition. The 74HC14 has Schmitt trigger inputs and the time delay is generated as C2 charges until the voltage on pin 11 drops below the gate threshold point.

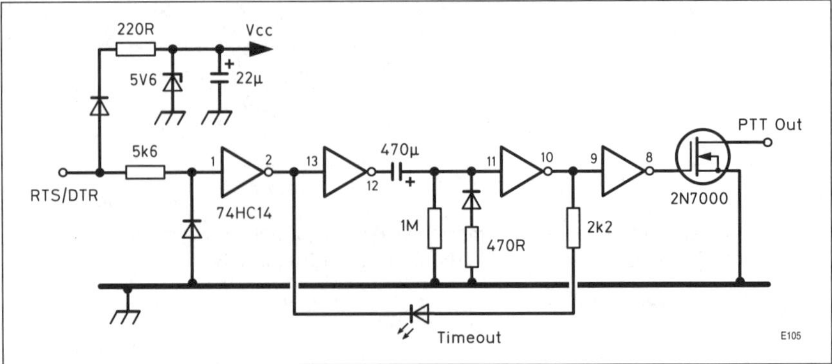

**Fig 3.6. Watchdog timer for transmitter protection**



The circuit is self-powered via diode D1 and the 5.6V zener, making use of the fact that the RS232 line can quite easily deliver several milliamps of current. For a longer timeout period the value of the 470μF capacitor can be increased, although new, 25V working devices should be used here to keep leakage current as low as possible. By also increasing the value of the 1MΩ resistor, even longer timeout periods can be achieved. The 470Ω resistor and D3 ensure a rapid discharge of C2 when returning to receive. The LED needs to be an ultra-high brightness type as only a couple of milliamps can be spared to operate it without excessively overloading the RS232 line.

## Automatic tuning and scanning

One way of making use of the transceiver control port is to implement a control panel on the host computer that allows memory functions, frequency shift working, fine tuning, and other functions to be made available that may not be present on the transceiver's front panel. A number of software packages exist for this purpose, both available for purchase (often from the transceiver manufacturer) and as freeware. Fig 3.7 shows the screen dump of the freeware programme Commander, available for download from reference [2]. This can cope with most makes and types of amateur receiver or transceiver, allows full frequency and mode control and implements 30 memories which contain all operating parameters, plus numerous other functions. Some manufacturers such as Ten-Tec make completely 'blind', ie front-panel-less, radios that make such a driver essential. The exclusion of manual controls such as tuning, volume and

**Fig 3.7. Commander rig control software**

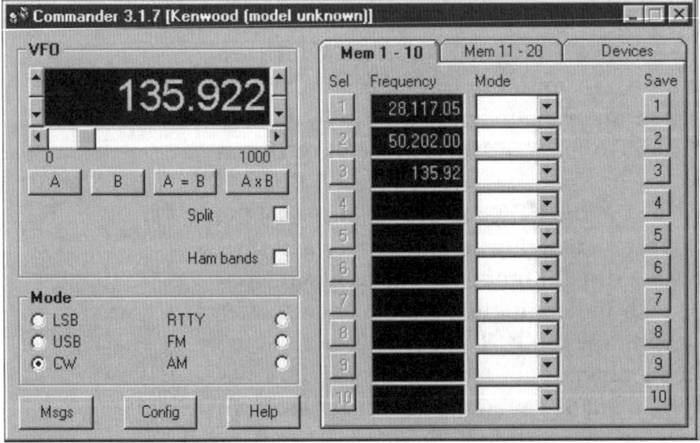

bandswitching reduces the cost of the hardware and lets users customise their interface for personal preferences.

## Spectral analysis

The next step on from single frequency control is to set up software to sweep, or monitor, bands of frequencies. These may be, for example, complete amateur bands to look for activity on an otherwise dead band, or on parts of bands looking for specific signals. A simple spectrum analyser can be generated by writing software that steps across a band of frequencies in fixed channels, then, by using the receiver's S-meter reading (this is often made available as a message back to the computer from the receiver), generates a plot of frequency versus signal strength. If signal strength information is not available as part of the software control commands, a separate analogue-to-digital converter can be used to monitor the AGC line or S-meter and feed its data back to the PC. A suitable A/D converter interface for this purpose is described in Chapter 4.

Fig 3.8. Screen dump of spectrum monitoring programme

Fig 3.8 shows the screen dump of a simple spectral analysis programme for the IC746.

Where operation is channelised, such as in the FM sections of the VHF amateur bands, band monitoring is made even simpler. The receiver is stepped across the channels and a single one-wire interface from the squelch circuitry is all that is needed to tell the software the channel is occupied. This single line can be interfaced to the PC via one of the unused RS232 handshake lines – such as CTS for example, or even by something as esoteric as a pin on the game port; all can be read by software.

## Algorithms and techniques for detecting signals in noise

Monitoring of beacons that are normally undetectable, but which can suddenly come up out of the noise, can be a rather time-consuming and boring task. It would be a lot easier if we could automatically detect the presence of a weak carrier in noise and either sound an alarm or start some recording software.

A simple threshold detector on the AGC line, or an audio filter followed by a tone detector as shown in Fig 3.9, is one means to detect signals that are strong enough to trigger the circuitry, but these techniques will not work for signals that are buried right down in the noise.

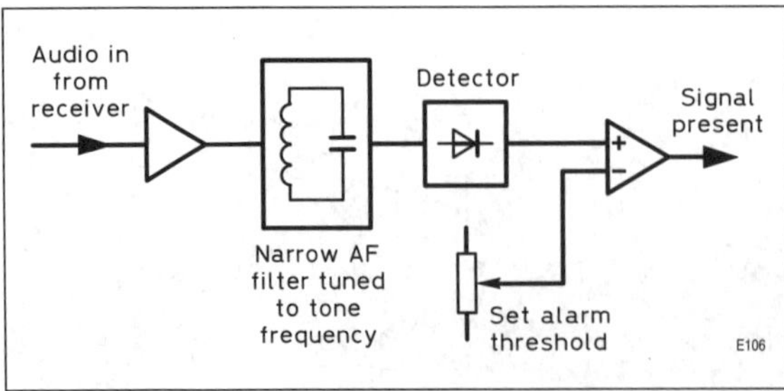


**Fig 3.9. Filter plus tone detector for automatic signal detection**

## Determine the frequency spectrum

A very sensitive carrier detection scheme can be devised by using some of the digital signal processing techniques that will be covered in Chapter 8, capable of generating a 'signal present' alarm at levels that can barely be heard by ear. The heart of the detection process uses the fast Fourier transform, a software technique covered in detail in Chapter 8, but for the purposes of this description it is easiest to visualise the process as a software-generated bank of band-pass filters. When a block of successive samples of a digitised audio input waveform are passed though the FFT process, the result is an array of numbers whose values are the amplitudes of each of the spectral components from DC to one half of the sampling rate. The bandwidth of each 'band-pass filter' term is defined by the block length. So, as an example, an audio waveform could be digitised at 10kHz and 512 samples passed into the FFT process at a time. The output would be 256 values of the spectrum from 0 to 5kHz, in units of 10000/512 Hz, or approximately 19Hz. Fig 3.10 shows this.

So we have filtered the input audio waveform from an SSB bandwidth down to around 19Hz, and more significantly we still have all the information in the original 3kHz available, so the need to accurately tune our signal to fit into a single 19Hz bandwidth filter has now disappeared. The rationale behind some of the numbers chosen for this example is covered in the FFT description later, but we should note that the FFT size has to be a power of two, such as 256, 512, 1024 etc.

### Effects of noise

We now have the array of numbers that indicate the signal power present in the 256 outputs of a bank of band-pass filters, each 19Hz wide, that cover our 0 to nominally 5kHz bandwidth – each one of these 256 values is termed a *frequency bin*, as in dustbin! This process is repeated at the block rate, so each set of 256 values is updated every 25.6ms. We now tune our receiver so the beacon frequency being monitored falls somewhere in the SSB pass band and, by monitoring the levels of each bin, when the signal appears, the value of the appropriate FFT bin corresponding to its tone frequency rises. If this exceeds a preset threshold an alarm can be sounded and whatever action to log the result taken. But now we run into problems with noise.

Noise is a random process and its frequency spectrum appears as many spikes at all frequencies, so in the absence of a real carrier, each FFT bin will be randomly outputting values, any one of which will be an instantaneous local maximum which will vary randomly from block to block and is very likely to exceed the triggering threshold at some time. This is where statistical analysis starts to play a part. A genuine carrier falling into one of the bins, even if it is only a fraction above the noise, will give an increased value for that bin but, if measured on its own (using one block's worth of data only), it is almost certain to be swamped by noise spikes that can easily be much bigger in many of the other bins, and a simple level detection will not give reliable results. The threshold can be raised to minimise these false alarms, but setting it too high means the signal has to be quite strong before it is detected.

## Noise averaging

The next stage is to average successive blocks of data. By calculating 256 separate moving averages for each of the bins over the last few blocks of data, we can build up a more accurate estimate of time-averaged energy in each of the bins. The averaged energy in each bin is now compared with the threshold and only if the signal in the same bin is high for several blocks will the alarm be sounded. This two-dimensional process, looking over frequency and time, goes some way to smoothing out the effects of noise and begins offer us a reliable automatic alarm when a signal appears. Visual spectral analysis programmes such as Spectrogram make use of time averaging in a similar way for detecting weak signal in noise, by making the human eye/brain combination look for the successive increased values in each bin. Greater reliability can be obtained by using a voting process. The signal in a bin has to exceed its threshold, say, three out of four times in a successive set of readings before an alarm is generated

However, what happens if the noise level changes, such as happens on HF during the day/night transition or if broad-band interference appears? If the noise level rises too much while the threshold remains unchanged, there is a reasonable probability that even averaged noise could trigger the alarm – so we need some way of moving the threshold automatically. In other words we want to measure the signal-to-noise ratio in each bin rather than the absolute level. A way to measure S/N ratio is to sum the entire signal power across all the bins – this total gives the power in the entire input signal of signal plus noise. By making the assumption that a signal, if it appears, will only be in one or two bins at the most, by dividing the total power by the number of bins we get an approximation to the noise level in each bin. Now the time-averaged power in each bin can be compared to the effective noise level and the signal-to-noise ratio for each bin calculated. The highest S/N ratio, if it exceeds a threshold, can be considered a valid signal detection.

There is another way of deriving the signal-to-noise ratio that does not require the calculation of the total power, is more immune to spikes and bursts of interference and is often computationally faster that summing all the powers in each bin. The technique relies on some of the statistical

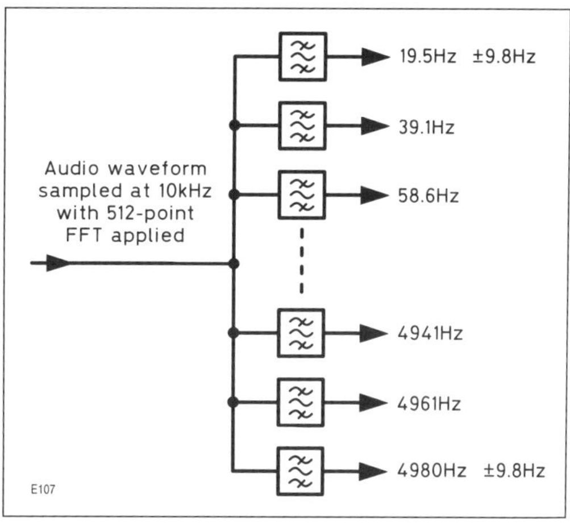


Fig 3.10. The FFT technique for generating a bank of effective band-pass filters

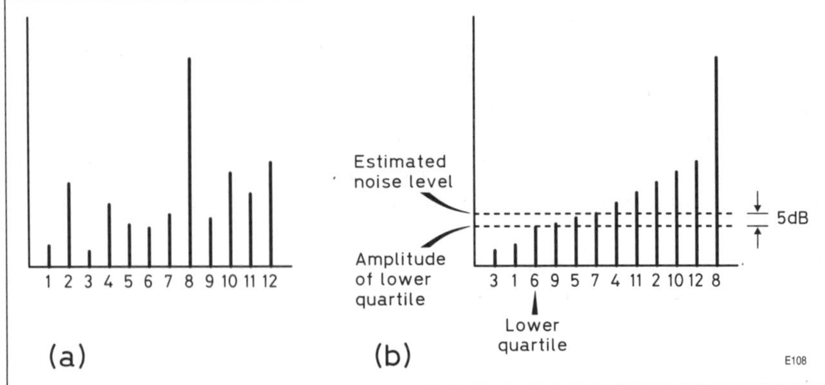



Fig 3.11. Generating the noise level by statistical methods. (a) Raw FFT output. (b) Bins re-ordered into ascending order of amplitude

properties of white (Gaussian) noise that we do not need to go into here in any detail. When the FFT process has generated its 256 values, these are sorted into ascending power level and the value represented by the lower quartile – the point where a quarter of the bins contain lower values; for a 256-point FFT, the 64th sorted bin is used. Statistical theory says that adding a value of 5dB to this gives the signal-to-noise ratio. Fig 3.11 shows this in more detail.

## Peak detector

One final aspect that needs to be covered is that of signal bandwidth. We have assumed so far that the wanted signal will fall in just one bin – but this is far from being a valid case. A pure carrier could easily sit at the crossover point of two bins, and each would give a valid answer. Furthermore, if the signal is modulated, even by keying sidebands, the modulation will spread into several adjacent bins. The solution is to detect peaks rather than single bin levels.

One peak detection technique is to step across each frequency bin in turn, comparing it with the previous one. If three successive rises are followed by a drop, then a valid peak has occurred on the previous bin. Fig 3.12 illustrates this peak-detecting algorithm. However, if the modulation is so wide that it spreads over many bins, this technique, which is really for detection of CW-type signals, is not really valid.

## Complete signal detection process

We now have a multistage algorithm that can give very reliable detection of low-signal-to-noise, CW-type signals in a varying environment with a low probability of false alarms. The process can be summarised:

1. Perform successive FFTs on blocks of data to generate the frequency spectrum of the input signal plus noise, and store these in an array.
2. Take the frequency/amplitude original data for the current block and sort it into ascending order of amplitude. Take the value represented by the lower quartile and add 5dB to give the noise level per bin. The original

spectral data needs to be preserved, so a separate copy of the data needs to be generated for the sort process.

3. For each block, step across each bin in the original (unsorted) data, looking for three rising values followed by a fall to indicate a peak. If this peak occurs at a signal-to-noise ratio greater than a preset threshold, then note the value of the bin where this peak occurs and store this value in an array containing the history of all peak detections from the last few blocks worth of data.

4. Compare the bin number of each valid peak detection with the peak detections of the last few blocks' worth of data using a voting system. If more than, say, three out of four peaks exceed the threshold successively in the same bin, then we have a very good assumption that a signal is present.

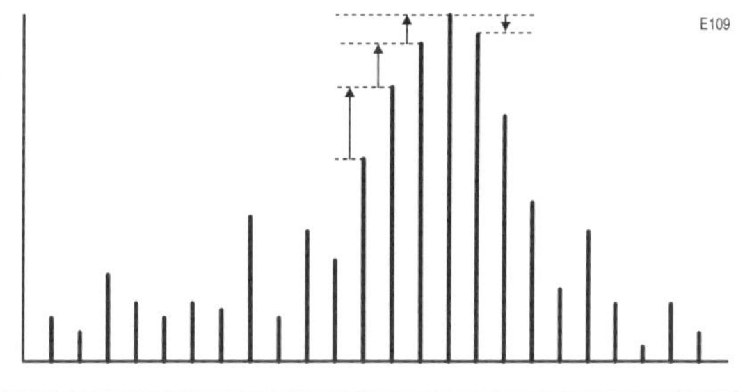

Fig 3.12. Peak detection by looking for successive rising amplitude levels, followed by a fall

This is a complex algorithm to design, but does give one of the most reliable automatic signal detection routines in existence. By fine tuning each of the variables, bin size, number of bins, peak detect rising and falling values, S/N threshold and number of valid votes, it can be made to work for many of the signal types likely to be encountered. The order of some of the steps can be swapped around, eg storing valid peaks in the history array before comparing to a threshold, or some stages can be left out – the easiest way is just to try it and see!

# IARU beacon monitoring

The IARU operates a chain of HF beacons across the world on single fixed frequencies in the 14, 18, 21, 24 and 28MHz bands. These beacons operate in sequence, each transmitting its callsign in CW for 10 seconds every three minutes, on each band in turn. The timings are controlled to extremely high accuracy by Global Positioning System satellite receivers, so at any instant it is possible to know exactly which beacon is transmitting on each band. BCNSCHED is a simple programme written by the author to allow easy monitoring of all these beacons; the PC's internal clock is first synchronised with UTC to the nearest second. Then, based on a table of the beacon timings, the software shows which call should be transmitting at that instant. When a complete session is complete on one frequency, the PC then switches the receiver to the next frequency and starts the listening session.

Although this programme was never actually finished, it could be enhanced to monitor the S-meter reading for each time/frequency interval and log the results to either a file or graphical display, allowing a map of HF propagation throughout the world to build up.

Fig 3.13 is a screen dump for this software showing how the data is presented to the user.

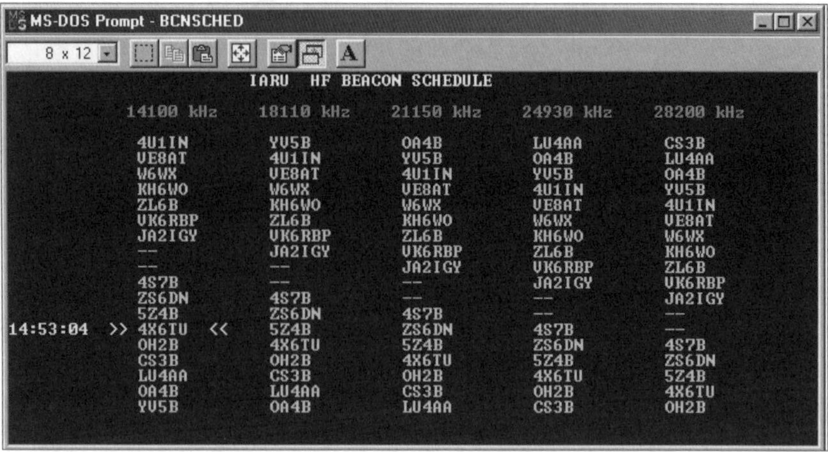

**Fig 3.13. BCNSCHED screen dump**

## Automatic link establishment

ALE is a process that has been used for some years by military and some commercial users of the HF spectrum, and involves automatically monitoring the path between a net of stations by sending and monitoring occasional probe signals transmitted by each station on each of the set of frequencies allocated to them. By measuring the strength of the received probe signals, which have of course to contain the identity of the transmitter sending each one, each station can build a continuously updated real-time map of propagation between itself and any station in the net. Thus, when communication is required the optimum frequency can be set automatically and the user can be insulated from the vagaries of HF propagation, choice of frequency, and sometimes even transmit power level; the HF link just appears as a normal voice channel.

The standards for this in the military world are now in the public domain and are encapsulated in a standard known as MIL STD 188-141A. Charles Brain, G4GUO, has written some software for the PC that fully implements this protocol by making use of the sound card to generate and decode the audio tones necessary to interface to an SSB transceiver for sending and reading the probe signal. Full control of the transceiver is possible, including transmit/receive switching, and tables can be set up of allowed frequencies which can be grouped for different nets. It is even possible to set receive-only permissions on certain frequencies, such as those allocated only to different ITU regions. A wide range of amateur transmitters and receivers can be controlled, and others can be added by Charles on request. Fig 3.14 shows a screen dump of the PCALE software in use; it can be freely downloaded from G4GUO's website [3].

One necessity for fully automatic transmit control is that of antenna tuning. The PCALE software can be set up to allow for antenna tuning time but this software, if it is used to transmit probe signals, presupposes an automatic ATU or broad-band transmit antenna is in use. A suitable automatic ATU is the Picatune by Peter Rhodes, G3XJP [4]. See Chapter 7.

# GPS

The GPS satellite system was mentioned in the previous section as an aid to getting very accurate time information. GPS receiver modules are now readily available from a number of manufacturers [5, 6] that give a simple ASCII text output of positional information on an RS232 interface, along with time and date and various other bits of navigational data such as speed, as well as satellite status. The time as sent on the serial interface is usually 'late' by a few hundreds of milliseconds, as it is sent after the event, but this time stamp is associated with another signal available separately which supplies one pulse per second, with a timing accuracy better than 1μs. In some cases and with some modules designed primarily for time keeping rather than navigation, this accuracy improves to a few tens of nanoseconds. Thus is now possible for any amateur to identify an exact epoch of time, and know that the same trigger is happening exactly at that instant anywhere in the world! This makes time-of-flight measurements of radio signals quite feasible and can allow ionospheric measurements to be made in real time. This aspect of GPS timing will be revisited in Chapter 7 where GPS is used for accurate frequency determination, but here we are just concerned with reading the data from the serial interface.

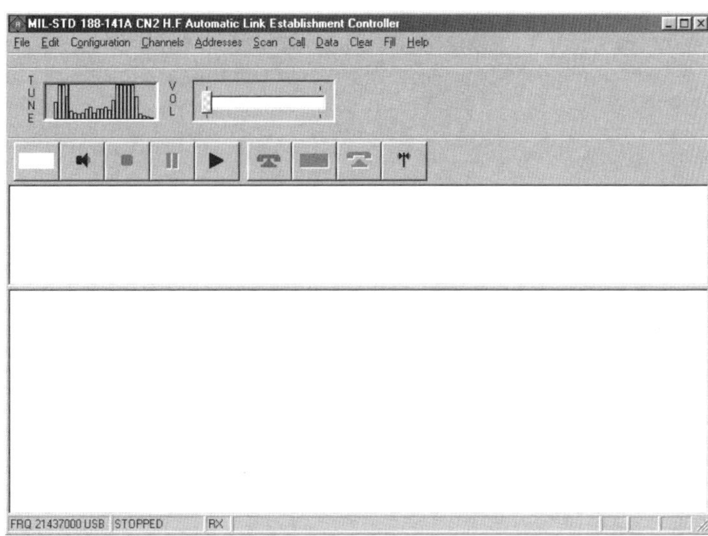

Fig 3.14. PCALE screen dump

The serial data is delivered in the NMEA0183 format, but is usually just referred to as *NMEA*, an abbreviation of 'National Marine Electronics Association'. This standard defines the data format and interfaces for all maritime navigational equipment, allowing for it to be freely interconnected. The NMEA specification defines messages from instruments in the form of plain text in a sentence structure, and all sentences associated with GPS start with the letters '$GP'. Many GPS receiver manufacturers also add a few extra sentences of their own for additional information, such as satellite status and housekeeping. A typical block of data sent over the RS232 link every second is shown in Table 3.2, exactly as it appears when the NMEA output is displayed using a terminal programme such as Hyperlink.

For our purposes, the first sentence beginning '$GPRMC' is the most useful – this is the NMEA Recommended Minimum Specific GPS (RMC) sentence. Data items are separated by commas as follows:

• The first item is the time, here 21:21:32 – this refers to the seconds pulse that has just happened.

**Table 3.2. Typical NMEA sentences from a GPS receiver module**

```
$GPRMC,212132,A,5054.5876,N,00117.4041,W,000.0,000.0,141202,003.5,W*7B
$GPGSA,A,3,,11,14,,28,31,,,,,,,3.7,2.4,2.7*38
$GPGSV,2,1,06,03,23,146,,11,64,276,40,14,33,083,44,20,21,215,36*74
```

- The 'A' indicates a valid position – if not present the previous time may be in error.
- The longitude in the format DDMM.MMMM, with leading zeros suppressed, here 50°, 54.5876' N.
- The latitude in the same format, here 1° 17.4041' W.
- Speed over ground, knots (zero).
- Course over ground in degrees.
- Date in the form DDMMYY, here 14/12/2002.
- Magnetic variation (3.5°W).
- A checksum, then the sentence is terminated with a carriage return linefeed pair, CR/LF.

The GSA sentence gives some of the data used in the position calculation and its accuracy. The GSV sentence lists the satellites in view. The NMEA standard specifies a baud rate of 4800, but other rates are possible from most GPS modules – a faster rate may be needed if more sentences are requested than can fit in a one-second burst at 4800 baud. Any software written to use the NMEA information sentence has first of all to detect the start by looking for the string "$GPRMC". It must then read in all data up to the CR/LF pair, then search for the positions of the commas. Once these are known the data can be extracted from between the commas using the knowledge of the sentence format.

# References


[1] 'In Practice', *RadCom* December 2002. The 'In Practice' website (www.ifwtech.co.uk) has many links to PC control interface circuits.
[2] Commander can be obtained from www.qsl.net/civ_commander/.
[3] G4GUO's website for PCALE, the MIL-STD 188-141A software, is www.chbrain.dircon.co.uk/.
[4] 'The PicATUne automatic antenna tuner', Peter Rhodes, G3XJP, *RadCom* September 2000.
[5] Garmin GPS25 module: www.garmin.com/products/gps25/.
[6] Motorola Oncore GPS Module: www.motorola.com/ies/GPS/pdfs/ m12.pdf.

# CHAPTER 4

# Microcontrollers

Microcontrollers are single chips that can probably be considered the simplest devices which can run stored programmes and execute actions based on the results. A microcontroller chip is usually a small processor that includes its own programme and data memory, a range of peripherals such as a serial interface and timers, and drivers for input and output pins; they are often designed to be used in completely stand-alone applications.

The first microcontroller chips were derived from microprocessors, first by including some on-chip memory and I/O drivers, then by adding various peripheral functions – devices based on these are in widespread use. For example, the original Motorola 6800 eight-bit microprocessor led to the 6805 microcontroller which is to be found in a number of automobile engine-management systems. Similarly, the 8080 family, which in one direction led to the PC, in another evolutionary branch became the 8051 controller, widely used in many industrial applications today. Both have a huge base of legacy software and applications, and for this reason will continue to be used for a long time to come. However, these traditional processor-derived controllers were not always the ideal solution for simple low-cost projects. It was time consuming to write software for them and to re-programme or update; furthermore the traditional memory architecture was not the fastest solution for high-speed operation. So, in the 'nineties a new generation of dedicated microcontrollers appeared. These made use of a very straightforward architecture and a limited instruction set, that of a *reduced instruction set computer* (RISC) processor. They also operated with separate programme and data memory areas to optimise speed. One of the most popular of these dedicated microcontrollers was the PIC, developed by Arizona Microchip, although there are other close runners, the Atmel AVR and the 8051.

## The Microchip PIC processor

The first PICs were very simple 18-pin chips with 13 programmable I/O lines and a simple RISC instruction set that consisted of just 32 commands. Other versions had more I/O lines in packages with up to 40 pins but were compatible in software. All PICs are made in CMOS technology with its consequential ultra-low power consumption when operated at low speed, ideal for battery applications, and can be used with a clock frequency up to 20MHz for more demanding tasks. The first PICs had 1k of 12-bit wide programme memory and a few tens of bytes of user memory – the exact figures depending on the actual chip type. By separating the programme and data memory in the so-called *Harvard architecture*, the 1k limit is actually a lot more use than it would be in a traditional sequential-memory architecture. Furthermore, dedicated programme memory means that for most instructions only (branches excepted) one clock cycle per instruction is needed. A 1MHz clock (generated on-chip from a 4MHz crystal) gives an instruction time of 1μs. Traditional architectures need, as a minimum, two and often more clock cycles per machine code instruction.

Originally, PICs were made using EPROM technology that could be re-programmed after erasure using ultra-violet light; low-cost applications used one-time programmable chips without the expensive quartz window and ceramic construction that is required for an erasable version. Soon afterwards, a new technology was introduced – the first EEPROM PIC appeared and could be electrically reprogrammed at will without having to use a UV light source. This offered the possibility of being reprogrammed *in situ*, ie while mounted in its target circuit. The instruction set was increased to 35 commands with a 14-bit word. The ubiquitous PIC P16C84 had appeared, which has been described as "the 741 of the microcontroller world" because of its universal adoption. An enhancement to the production process, using Flash technology to reduce cost and improve reliability, led to the 16F84 which is (very nearly) functionally identical to the P16C84 and is still the standard workhorse of the PIC world. This has 1k of programme memory and 68 bytes of data memory, 13 I/O pins, an internal timer with prescaler, 64 bytes of user non-volatile memory, and four interrupt sources. It will work over the power supply range 2 to 6V with clock speeds of up to 20MHz on selected devices, and down to DC for low-power operation. At the lowest clock speeds, current consumption is just a couple of milliamps, and in sleep mode even lower, at a few microamps. The instruction set for the P16xxx series of PICs is shown is Table 4.1, and a simplified diagram of the architecture of the 16F84 device in Fig 4.1. Full data sheets and information are available from the Microchip website [1].

## Programming the PIC

Programming a PIC for any task is similar to writing in any other programming language, although much more knowledge of the target processor is needed. PIC programming in high-level languages such as C or Basic is possible, and many users prefer this route. The best compilers can hide the complexities of the chip but these versions can be quite expensive – low-cost, and even free or shareware, C language compilers do exist, but using these often gives little advantage over assembler.

There is no doubt that for optimum processing speed and keeping

## Table 4.1. PIC 16Fxxx instruction set

| Mnemonic/ operands | Description | Cycles | Status bits affected | Notes |
|---|---|---|---|---|
| **Byte-oriented file register commands** | | | | |
| ADDWF $f$, d | Add W to $f$, place result in d ($f$ or W) | 1 | C, DC, Z | 1, 2 |
| ANDWF $f$, d | AND W with $f$, place result in d ($f$ or W) | 1 | Z | 1, 2 |
| CLRF $f$ | Clear $f$ | 1 | Z | 2 |
| CLRW | Clear W | 1 | Z | |
| COMF $f$, d | Complement $f$, place result in d | 1 | Z | 1, 2 |
| DECF $f$, d | Decrement $f$ | 1 | Z | 1, 2 |
| DECFSZ $f$, d | Decrement $f$, skip next instruction if zero | 1 or 2 | | 1, 2, 3 |
| INCF $f$, d | Increment $f$ | 1 | Z | 1, 2 |
| INCFSZ $f$, d | Increment $f$, skip next instruction if zero | 1 or 2 | | 1, 2, 3 |
| IORWF $f$, d | Inclusive OR W with $f$ | 1 | Z | 1, 2 |
| MOVF $f$, d | Move $f$ to destination | 1 | Z | 1, 2 |
| MOVWF $f$,d | Move W to $f$ | 1 | | |
| NOP | No Operation | 1 | | |
| RLF $f$, d | Rotate left $f$ through carry | 1 | C | 1, 2 |
| RRF $f$, d | Rotate right $f$ through carry | 1 | C | 1, 2 |
| SUBWF $f$, d | Subtract W from $f$ | 1 | C, DC, Z | 1, 2 |
| SWAPF $f$, d | Swap nibbles in $f$ | 1 | | 1, 2 |
| XORWF $f$, d | Exclusive OR W with $f$ | 1 | Z | 1, 2 |
| **Bit-oriented file register commands** | | | | |
| BCF $f$, b | Clear bit B of register $f$ | 1 | | 1, 2 |
| BSF $f$, b | Bit set $f$ | 1 | | 1, 2 |
| BTFSC $f$, b | Bit test $f$, skip next instruction if clear | 1 or 2 | | 3 |
| BTFSS $f$, b | Bit test $f$, skip next instruction if set | 1 or 2 | | 3 |
| **Literal and control operations** | | | | |
| ADDLW k | Add literal to W | 1 | C, DC, Z | |
| ANDLW k | AND literal with W | 1 | Z | |
| CALL k | Call subroutine | 2 | | |
| CLRWDT | Clear watchdog timer | 1 | | |
| GOTO k | Go to address | 2 | | |
| IORLW k | Inclusive OR literal with W | 1 | Z | |
| MOVLW k | Move literal to W | 1 | | |
| RETFIE | Return from interrupt | 2 | | |
| RETLW k | Return with literal in W | 2 | | |
| RETURN | Return from subroutine | 2 | | |
| SLEEP | GO into standby/sleep mode | 1 | | |
| SUBLW k | Subtract W from literal | 1 | C, DC, Z | |
| XORLW k | Exclusive OR literal with W | 1 | Z | |

**Notes**

1. When an I/O register is modified as a function of itself (eg MOVF PORTB, $f$) the value used will be that value present on the pins themselves rather than that stored in the output registers.
2. Clears prescaler when executed on the TMR0 register
3. If the Programme Counter (PC) is modified or a conditional test is true, the instruction requires two cycles. The second cycle is executed as a NOP.

resultant code size small, as well as understanding all the device's idiosyncrasies, the native assembler is much more efficient than a high-level language such as C. For some tasks, however, such as string handling, a high-level language can prove more suitable.

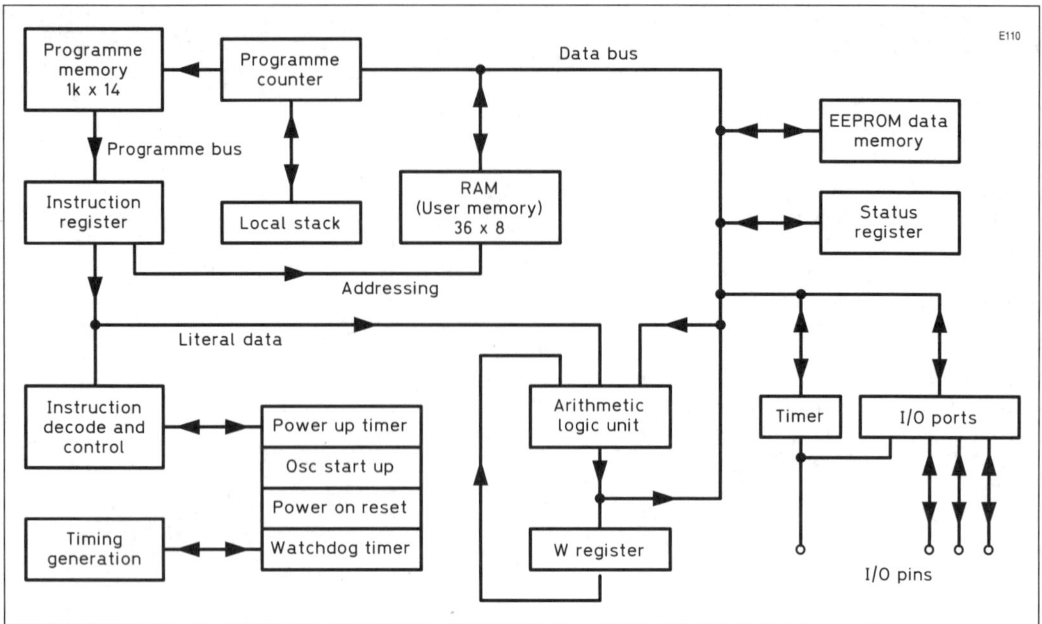


**Fig 4.1. Simplified architecture for the baseline PIC controller**

To develop PIC software (more correctly called *firmware* as it is embedded into a final application) a personal computer of some sort is invariably used to develop the code using various software tools supplied, usually free of charge, from the PIC manufacturers. A programmer driven from the PC is then used to load the developed code into the chip. Programmers of different sorts and levels of sophistication can either be purchased from most electronic component suppliers, or simpler ones capable of programming many, but not necessarily all, of the PIC processor types can be quite easily built. Most programmers run off either the parallel port or the serial port and some of the commercially manufactured units are now offered with a USB interface. Many popular low-cost devices are restricted to programming the 16F84 device.

There are four stages to producing a fully functional PIC design:

1. Design the circuit and hardware, making use of the PIC data sheets to determine which pins are to be used for each function – many of the special peripheral functions require their interfaces to be on specific pins.

2. Write the assembler code – the list of instructions to perform the function required. This is in the form of an assembler (source) file in text format and usually generates a file with a name like xxx.ASM where xxx is the filename. There is usually a section that defines memory locations and register names and sets up peripheral functions, followed by the operating code itself. A range of INCLUDE files are provided by Microchip for each processor type to make naming programming easier by naming all the registers and individual control bits. On a PC, a text editor such as Notepad or the DOS Edit command can be used to write the source code.

| Table 4.2. PIC programmers | |
| --- | --- |
| PICSTART Plus | Made by Microchip and available from many suppliers. Uses the serial port for interfacing and is driven through Microchip's MPLAB software which is a full PIC development suite available for the Windows operating system, including editor and simulation packages. |
| EPIC | Supplied by MELabs [2]. A simple programmer supplied as a PCB with ZIF socket, making use of the parallel port. |
| WISP628 | A design for home construction that requires a programmed PIC as part of the design. A chicken-and-egg situation arises here if this is to be the first programmer, but ready-programmed PICs are available from the designer in reference [3]. |

3. Assemble the source code into machine code. The assembler generates a file xxx.HEX which contains the machine code in a form useable by the programmer. It also generates a full listing xxx.LST which contains a repeat of the source code with all associated memory maps, variables and labels shown. Any errors in the source code such as illegal calls, mistyped labels or variable names or illegal values appear in an error file xxx.ERR; this will be empty for a 'perfect' assembly. Warnings are also produced when certain operations that may cause potential problems in some circumstances are made. An excellent assembler is MPASM, produced by Microchip, which can be downloaded from reference [1].

   MPASM is called from the command line by invoking it along with the filename of the .ASM file, eg MPASM TESTPROG. After assembly is complete, it reports back on the number of errors encountered in the process along with any messages or warnings about the code. Details of these are stored in the .ERR file.

4. Programme the chip. Most modern PICs are programmed serially by making use of three of the pins plus ground. The master reset pin is raised to +13V while the chip is powered normally from its 5V rail, and the data is clocked in serially using two pins, one for clock and the other for data. Programmers usually include a zero insertion force socket for programming raw chips, or provide the four connections needed for in-circuit programming. Microchip supply a range of programmers and debugging tools, but details of some low-cost ones suitable for home constructors are included in Table 4.2. This list is by no means exclusive and many more PIC programmer designs can be found on the web.

An intermediate stage between steps 3 and 4 is available – that of simulation. Host computer software takes the .ASM or .HEX files and simulates the functioning of the PIC hardware with the user code which can be single stepped if required, allowing all intermediate values, the state of all registers and I/O lines to be examined at any time for debugging. Alternatively, breakpoints can be set and registers checked at this point.

It is only a simulation, and the functioning of peripherals such as A/D converters has to be assumed. Timing and clock cycle count is shown

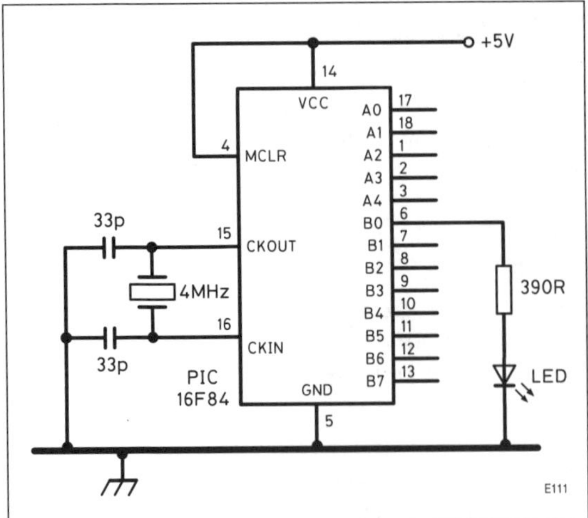



**Fig 4.2. Circuit diagram for PIC based LED flasher**

directly, a function which can be very tedious if undertaken manually on the source code.

## LED flasher programme for the 16F84 PIC

As an introduction to PIC programming, a simple PIC programme is given in Table 4.3 that flashes a LED. This is written for the 16F84 processor but will also run unmodified on many 16xxx types, although memory locations may have to be changed for later PIC devices with more peripheral functions on-chip that use up the lower memory positions. Note that this programme will not run on the original baseline 16C5x family devices with their 12-bit architecture without modification.

All text after a semicolon is a comment, is not needed as part of the programme and is ignored by the assembler. The circuit diagram is shown in Fig 4.2.

*Detailed description of the PIC assembly code for the LED flasher*
The first section defines the processor type to the assembler and calls up the INCLUDE files that define all the internal registers and control bits. Next, data memory locations are defined as variables with meaningful names. The CBLOCK command starts these allocations at address 0x0C as locations below this are taken up by the special function registers, such as the I/O ports and status register.

The 16F84 PIC has 13 I/O lines, five on Port A and eight on Port B. Here we connect the LED to the lowest bit on Port B, referred to as 'B0', or PORTB, 0 as shown in the listing. To make life easier, and to illustrate good programming practice, a #DEFINE statement is used so that later references can just call up the LED pin by name. If it is decided to subsequently change the pin that the LED is connected to, all that is needed is one change to this #DEFINE statement rather than searching through the code for every occurrence of PORTB, 0 and changing it. Within many of the listings and subroutines in this book, I/O lines and some other functions will usually appear as meaningful names without any corresponding #DEFINE being shown in the listing. Whenever this is encountered, a corresponding #DEFINE will have to be included at the start of the programme to define whichever port/pin combination is to be used.

ORG 0 defines the start of the code at the beginning of programme memory, and here we use a trick which is good programming practice, although not essential. The interrupts on this family of PIC devices always causes a programme jump to memory location 004, just after the beginning. If interrupts are not being used and are disabled, this location can just be part of

## Table 4.3. PIC programme to flash an LED

```
;LEDFLASH.ASM
    list    P=PIC16F84      ;Define processor type to assembler
    include P16F84.INC      ;Include file for special function register and bit names
    cblock  0x0C            ;Start of accessible data memory above special function registers
        DelCount1           ;Two variables used in the delay routine
        DelCount2
    endc
    #define     LED             PORTB, 0   ;Good practice and makes things easier later
;------------------
    org     0               ;Start of programme memory
    clrf    INTCON          ;Disable all interrupts by clearing interrupt control register
    goto    StartUp
    org     0x04            ;Interrupt vector
    retfie                  ;Just in case of accidental interrupt

StartUp
    bsf     STATUS , RP0 ;Data memory Page 1
    movlw   b'11111110'  ;Shown in Binary format
    movwf   TRISB        ;B0 as output for LED, rest inputs
    bcf     STATUS , RP0 ;Back to Page 0
MainLoop
    bsf     LED
    movlw   d'20'           ;On for 20ms,  number defined in decimal format
    call    Delay
    bcf     LED
    movlw   d'250'
    call    Delay
    movlw   d'250'
    call    Delay
    movlw   d'250'
    call    Delay
    movlw   d'230'          ;Off for 980ms
    call    Delay

    goto    MainLoop


;==============================
Delay ;Delay W ms (approx) assuming 4MHz clock
    movwf   DelCount2
DelLoop2                    ;Outer Loop
    movlw   d'200'
    movwf   DelCount1
DelLoop1                    ;Inner loop 5N-1  + 2 to set up = 5.N + 1
    nop
    nop
    decfsz  DelCount1
    goto    DelLoop1        ;End of inner loop

    decfsz  DelCount2
    goto    DelLoop2        ;End of outer loop

    return
;------------------
    end
```

the normal programme flow. *However,* if by accident or programming error an interrupt is generated at any time and for any reason, a jump to this location will occur – usually with unpredictable or fatal results. Furthermore, as this is the result of an error which may be completely remote in the code, its cause will be very difficult or even impossible to find. So, at location 004 we include a RETFIE command which immediately returns from any interrupt back to the correct place in the calling routine, causing no damage. Normal programme flow misses the interrupt vector and its return command by virtue of the GOTO StartUp command at programme memory location 000.

At StartUp, the direction of the ports is now defined. By default they are set as inputs and, as we want B0 to be an output, we need to store a '0' at the appropriate location in the Port B direction register, TRISB. Binary nomenclature b'11111110' is used here to make the bit pattern obvious on inspection, but it could equally well be defined in hex (0xFE or h'FE') or even in decimal (d'254). Data memory in this family of PICs is organised into banks, and the lesser-used registers are placed in Bank 1 with the common ones in Bank 0. As the TRIS registers are in Bank 1 this has to be defined by the statement BSF STATUS, RP0 which sets the appropriate bit in the Status register. It is cleared after TRISB is set. No other special function registers need to be set up for this programme so we can jump straight into the code to flash the LED.

A label, MainLoop, defines the start of this part of the code. The LED is first turned on with the BSF LED, or bit set command. Remember that the #DEFINE statement earlier really means that this translates to BSF PORTB, 0. The W register, which can be thought of as a working accumulator, is then loaded with a constant value of 20 (decimal) to define the length of delay required, ie the duration the LED is turned on for. The delay subroutine (described later) is called, which holds up the main programme flow for approximately the number of milliseconds passed to it in the W register; when this is complete the software returns to the next command after the call instruction. The LED is now turned off with the BCF command (bit clear register F) and W loaded with the value to define the off time, here decimal 200. As we are using an eight-bit data word, the maximum value here is 255; we need a longer delay than 255ms so the call is repeated three times to make up a total of 980ms off time. The programme now jumps back to Mainloop and the process continues indefinitely, giving a 20ms flash every second.

The Delay routine works as follows. The time delay is generated by putting the software into a loop with a controlled number of clock cycles. Assuming a 4MHz crystal, each instruction takes exactly 1µs apart from jumps, and some others that change programme flow, which take two clock cycles. There are two nested loops to generate potentially quite long delays – the inner loop starts at the first label, Loop1, and loads a constant into W which is then placed into the variable DelCount1. The DECFSZ command decrements this counter and tests to see if the result is zero – if so it skips the next instruction, otherwise it just continues. So, if the result is not zero the GOTO command is executed which completes the inner loop. This loop is made exactly five clock cycles long by including two no-operation commands to extend the overall length to this value. The final jump out of the loop loses a spare clock cycle, but there are two taken up with the initial load of the constant, so the total length of the inner loop is $5.N + 1$ where $N$ is the initial value loaded into the DelCount1 variable. Here a value of 200 decimal is used so the inner loop completes in close to 1ms.

The outer loop works the same way, but the value passed into the Delay routine is loaded into the DelCount2 variable, so the inner loop with its 1ms delay is executed this number of times. On completion the RETURN command passes programme flow back to the calling code, after generating a delay (nearly) equal to the value passed into it via the W register. The END statement is self-explanatory, and necessary! If more accurate timing is wanted, more attention has to be paid to the delay routines and it is here that simulation proves its value. However, the best way of obtaining accurate delays is to use the internal timer to generate an interrupt irrespective of code length, then rewrite the entire programme on an interrupt-driven basis. Some examples of this method of accurate timing are included later.

Little tricks like using binary or decimal nomenclature for numbers, adding numerous comments, indenting and separating sections of code with spaces or lines makes producing PIC assembler code that works first time easier; it certainly makes debugging simpler.

## Interfacing to PCs – a serial interface

One of the many useful functions for PICs in the amateur radio world is to simplify input and output from PCs for data logging and control purposes. In Chapter 3 we showed how PC interfacing could be achieved through the parallel and serial ports. PICs can be used to convert these formatted signals into real-world control and data, for example driving relays or reading voltage levels.

The listing given in Table 4.3, together with the circuit of Fig 4.3, shows a 16F84-based design to read an eight-bit byte sent over an RS232 link, decode this data and drive eight data lines, one corresponding to each bit of the data. These in turn drive buffers for relays. For error checking the data is returned over the RS232 link so the PC can compare that sent with the final state of the latches. A LED is flashed for each byte received to assist in debugging. As the 16F84 has no internal circuitry for handling serial data, this task is managed in software. On receive the programme sits in a loop looking for the '1' to '0' transition associated with a start bit but, as this circuit interfaces directly to the RS232 signal, voltage polarities are opposite to their logical equivalents, and the 'unofficial' 0/5V levels are adopted on the RS232 link. When the edge is detected, the software waits for half a baud period so the sampling instant is located at the middle of the data bit. The signal state is then checked again to make sure that the start bit is still valid, and if so the eight data bits are read by counting one baud intervals then shifting the result into the receiving register. After eight bits have been received, the data is

**Table 4.3. RS232 relay driver PIC source code**

```
;        RELAYDRV.ASM
;     Receives byte on RS232 line at 1200 baud, outputs data to Port B,
;     flashes LED then returns data on RS232 line


        LIST P=16F84

        __config    0x3FF1   ;Set crystal oscillator, WDT off, PUT on
        INCLUDE "p16F84.inc"
        cblock 0x0C
            BitCount
            DelCount
            DelCount1
            DelCount2
            Temp
            Counter
        endc


        #define     RXD     PORTA , 0   ;Serial data in from PC
        #define     TXD     PORTA , 1   ;Echoed data / status out to PC
        #define     LED     PORTA , 2

org     0
        nop
        clrw
        movwf   INTCON          ;disable interrupts
        goto    startup         ;jump to main code
        retfie                  ;Interrupt vector, Just in case
startup
        bsf     STATUS,RP0      ;ram page 1
        movlw   b'00000001'     ;
        movwf   TRISA           ;set port A0 as input for RS232, rest outputs
        movlw   b'00000000'     ;All outputs for relay data
        movwf   TRISB           ;
        bcf     STATUS,RP0      ;ram page 0
        clrf    PORTB           ;Start with all off
;--------------------------------------------
MainLoop
        call    WaitData ;Wait for character on RS232 line
        movwf   PORTB    ;Place received character on relay driver port
        bsf     LED      ;Flash LED for duration of returned character
        movf    PORTB, W ;Get relay data to W
        call    Send232  ;Send data back to PC for checking
        bcf     LED      ;Vey quick flash
        goto    MainLoop ;Continue indefinitely
;============================================================
BitDelay    ;total incl call and return, (3N+5) = 44us
        movlw   d'13'    ;+8 used in Tx/Rx routines = 52us
        movwf   DelCount ;  for 19200 baud
LoopBit
        decfsz  DelCount
        goto    LoopBit
        return
;------------
WaitData
```

**Table 4.3 (continued)**

```
        btfss    RXD        ;look for start transition, low/high = 1/0
        goto     WaitData
        movlw    d'7'
        movwf    DelCount
        nop
LoopStartBit        ;this loop 3.N - 1 long = 17
        decfsz   DelCount
        goto     LoopStartBit
        btfss    RXD        ;make sure it's still high, now centre of bit
        goto     WaitData
        movlw    8
        movwf    Counter
        clrf     Temp       ;26us to here from transition = half bit
ByteLoop
        call     BitDelay
        nop
        bcf      STATUS , C
        btfss    RXD
        bsf      STATUS , C
        rrf      Temp
        decfsz   Counter
        goto     ByteLoop ;this loop BitDelay + 8
        movf     Temp , w ;Returns with stop bit+  slack period
        return      ;received byte in Temp and W
;──────────────
Send232     ;Enter with data in W, send it to PC
        movwf    Temp
        bsf      TXD        ;start bit
        call     BitDelay ;44us, need 52 total
        movlw    8
        movwf    Counter
        nop
        nop
        nop
SendDatLoop
        rrf      Temp       ;bit into C
        btfsc    STATUS , C
        bcf      TXD
        btfss    STATUS , C
        bsf      TXD        ;could have 2us jitter depending on 1/0
        call     BitDelay
        decfsz   Counter
        goto     SendDatLoop      ;loop 8 + BitDelay long
        bcf      TXD        ;stop bit
        call     BitDelay
        nop
        nop
        nop
        nop
        nop
        nop
        return
;──────────────
        end
```

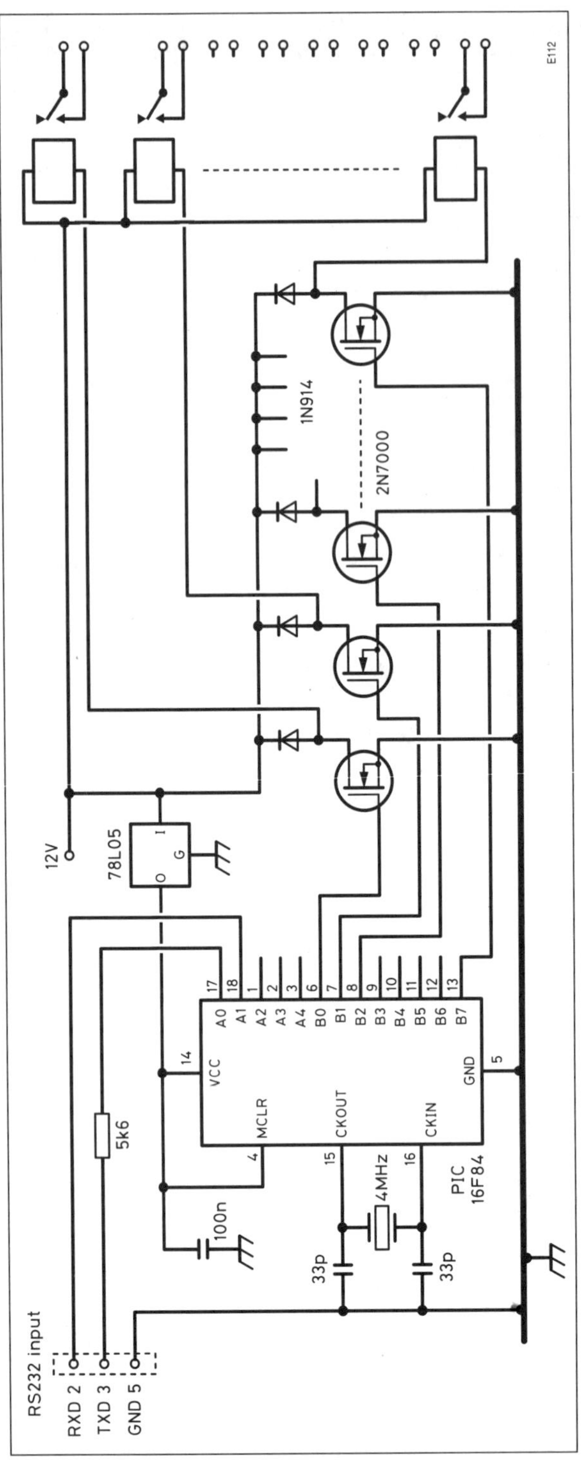


**Fig 4.3. RS232 controlled relay driver**



then sent directly to Port B to drive the relays. For checking, the port register is read then the result is encoded by a shift and bit set/reset loop timed at one baud duration. At 19200 baud, timing is quite critical so machine cycles for the RS232 send and receive routines have been tailored to be accurate to the nearest two cycles (2µs) which is acceptable. For higher reliability, a lower baud rate could have been chosen, and the `BitDelay` routine modified accordingly.

The routines `Send232` and `WaitData` occur within many programmes in this book whenever an RS232 link is required, and will not always be shown in listings to save space. In most cases the variables associated with these routines will have the same names but the pins and port bit allocations may change. Many of the more-modern PIC devices, such as the 16F628, have universal synchronous/asynchronous receiver transmitter (USART) hardware for handling RS232 links, and for other serial formats. A dedicated USART makes programming a lot easier, particularly as the processor can be doing other jobs while waiting for the serial characters to arrive instead of sitting inside an endless loop just waiting for data. On the downside though, baud rates are fixed to a few standard values, the I/O pins allocated to the serial interface are usually fixed and signal polarity is such that an RS232 driver chip, or at least a logic inverter, is necessary to interface to the usual serial port.

## Analogue/digital conversion

Some PIC devices such as the P16C71 have an internal eight-bit A/D

**Table 4.4. Some serial A/D converter chips suitable for interfacing to PICs**

| Type number | Pins | Number of bits | Number of channels | Max sampling rate (samples/s) | Interface |
| --- | --- | --- | --- | --- | --- |
| TLC549 | 8 | 8 | 1 | 45k | 3-wire Clock, Data, CS |
| MCP3201 | 8 | 12 | 1 | 100k | 3-wire Clock, Data, CS |
| MCP3202 | 8 | 12 | 2 | 100k | 4-wire, Clock, Din, Dout, CS |
| MCP3204 | 14 | 12 | 4 | 100k | 4-wire |
| MCP3208 | 16 | 12 | 8 | 100k | 4 wire |
| LTC1286 | 8 | 12 | 1 (differential) | 12.5k | 3-wire |
| LTC1298 | 8 | 12 | 2 | 11.1k | 4-wire |
| MAX1236/7 | 8 | 12 | 4 | 100k | 2-wire I2C |
| MAX1238/9 | 16 | 12 | 12 | 100k | 2-wire I2C |

converter included. This particular chip can accept up to four analogue inputs, one of which can be measured at any time by addressing an internal analogue multiplexer. The A/D reference may be provided either externally on one of the analogue input pins, or the +5V rail internally may be used as a reference. Other chips such as the 16C710 also have A/D converters but with slightly different characteristics – some recent devices have higher-resolution 10-bit converters and the relevant data sheet should be consulted for the exact operation in each case.

Instead of these PIC devices, an external A/D converter chip can be connected to provide functions not available on the devices with internal converters such as a higher resolution or more analogue input lines. There are a great number of suitable serially controlled devices which interface with their host processor through two, three or sometimes four signal lines, typically data, clock and strobe; several devices make use of a two-wire I2C bus which is described later.

Another advantage of using an external A/D converter is that software can be written for the Flash reprogrammable 16F84 which does not have its own A/D converter, saving the need for an EEPROM eraser every time the chip is to be reprogrammed. Some suitable serial A/D converter chips are listed in Table 4.4.

### Serial port A/D interface

The circuit shown in Fig 4.4 is for an interface that samples an analogue signal using the internal A/D converter of a PIC 16C71, then outputs the binary data on a serial RS232 interface. Although originally designed for digital signal processing and sampling of audio waveforms, it can be used for measurements down to DC.

The signal can be sampled at two rates of 100µs or 140µs between measurements. For DSP purposes the accuracy of the sampling rate is critical, and this is set by fine tuning the PIC code to give exactly the rate required. At 100µs sampling, 10000 eight-bit samples are generated each second and these have to be sent at a suitable rate on the RS232 interface. The fastest rate that most PCs will accept is 115200 baud so, by adopting a format of one start, eight data and two stop bits, 11 baud periods per sample are

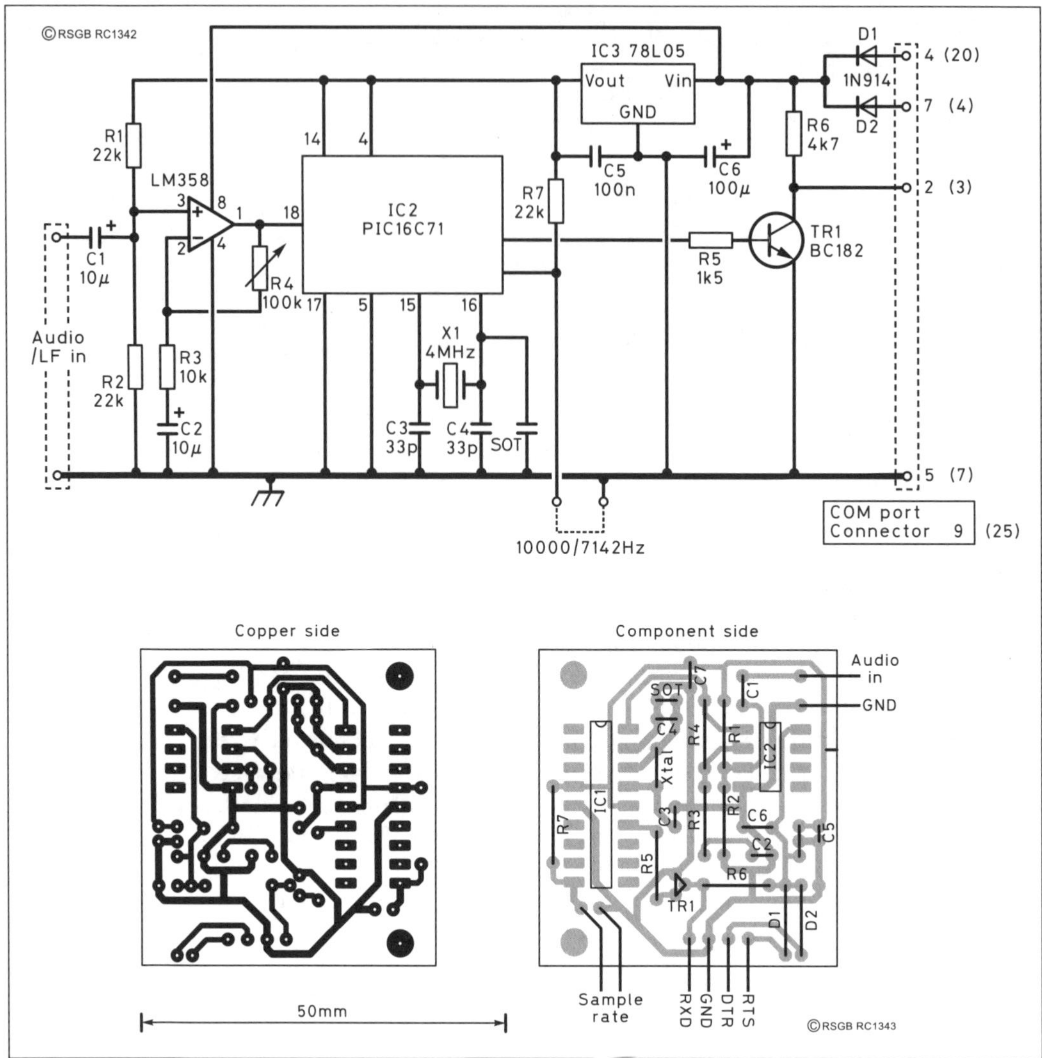



**Fig 4.4. Serial port A/D interface**

required, allowing a maximum data rate of $115200/11 = 10472$ bytes per second. Hence a sampling rate of 10000Hz is feasible with a small amount of timing or processing overhead.

The source code listing is shown in full in Table 4.5 and was originally written by Lee Wiltshire, G0IAY. At this high data rate we cannot afford the luxury of calling dedicated RS232 subroutines as there is very little spare time available for the calls and returns.

Instead, the serial bits are hard coded in one single loop that is exactly 100 clock cycles long in total to give a 10000Hz sampling rate. The internal A/D converter needs at least 85 clock cycles for its internal operation so, instead of triggering a conversion then waiting for it to complete, the data sent is that of the last conversion.

So the process now becomes:

**Table 4.5. Full source code listing for the analogue-to-serial Interface**

```
;Software to sample data at RA1 at 10kHz, and output it serially at 115.2kbaud to a PC
;Original written by Lee Wiltshire, G0IAY
LIST P=16C71
INCLUDE "P16C71.INC"
;port A is used as follows: bit 0 - not used (tie to ground
;                           bit 1 - analogue input
;                           bit 2 - not used
;                           bit 3 - Vref
;port B is used:    bit 7 - RS232 output (TTL Low = RS232 +ve using interface)
;                           bits 6 - 0 not used (set to inputs)
cblock 0x0C
    LAST_RDG        ;Reading from LAST A/D conversion
    TEMP
endc

org     0
    nop
    clrw
    movwf    INTCON         ;disable interrupts
    goto     startup        ;jump to main code
    retfie          ;interrupt service routine code

startup
    bsf      STATUS,RP0        ;ram page 1
    movlw    b'11111111'      ;
    movwf    TRISA            ;set port a as inputs
    movlw    b'01111111'      ;set Port B:7 as output, PB6-0 input
    movwf    PORTB
    bcf      STATUS,RP0       ;ram page 0
    movlw    b'10000000'      ;set RS232 to quiescent state
    movwf    PORTB

;setup the A-D here
    movlw    b'01001001'      ;Tad = fosc/8, ch1, A/D on
    movwf    ADCON0
    bsf      STATUS,RPU
    movlw    b'00000010'      ;set RA0,1 as analogue, RA2,3 as digital
    movwf    ADCON1
    bcf      STATUS,RP0
    clrf     LAST_RDG
loop                         ;Entire loop should be exactly 100 cycles long
    bsf      ADCON0,GO        ;set conversion going whilst we send last sample
    movlw    b'00000000'      ;1 start bit
    movwf    PORTB           ;bit time is 8.67us, so we will have +/- 0.5us jitter
    nop                      ;this bit 9us
    nop
    nop
    nop
    clrf     TEMP            ;clear output reg (not absolutely necessary)
    rrf      LAST_RDG,1      ;rotate bottom bit to carry
    rrf      TEMP,1          ;rotate carry into top bit
    movf     TEMP,0          ;get ready to output
    movwf    PORTB           ;and output bit
    nop                      ;this bit 9us (bit 0)
```

**Table 4.5** *(continued)*

```
        nop
        nop
        nop
        clrf    TEMP
        rrf     LAST_RDG,1
        rrf     TEMP,1
        movf    TEMP,0
        movwf   PORTB
        nop                     ;this bit 8us (bit 1)
        nop
        nop
        clrf    TEMP
        rrf     LAST_RDG,1
        rrf     TEMP,1
        movf    TEMP,0
        movwf   PORTB
        nop                     ;this bit 9us (bit 2)
        nop
        nop
        nop
        clrf    TEMP
        rrf     LAST_RDG,1
        rrf     TEMP,1
        movf    TEMP,0
        movwf   PORTB
        nop                     ;this bit 9us (bit 3)
        nop
        nop
        nop
        clrf    TEMP
        rrf     LAST_RDG,1
        rrf     TEMP,1
        movf    TEMP,0
        movwf   PORTB
        nop                     ;this bit 8us (bit 4)
        nop
        nop
        clrf    TEMP
        rrf     LAST_RDG,1
        rrf     TEMP,1
        movf    TEMP,0
        movwf   PORTB
        nop                     ;this bit 9us (bit 5)
        nop
        nop
        nop
        clrf    TEMP
        rrf     LAST_RDG,1
        rrf     TEMP,1
        movf    TEMP,0
        movwf   PORTB
        nop                     ;this bit 9us (bit 6)
        nop
```

**Table 4.5 (continued)**

```
        nop
        nop
        clrf    TEMP
        rrf     LAST_RDG,1
        rrf     TEMP,1
        movf    TEMP,0
        movwf   PORTB
        nop                     ;this bit 8us (bit 7)
        nop
        clrf    TEMP
        rrf     LAST_RDG,1
        rrf     TEMP,1
        movf    TEMP,0
        movlw   b'10000000'
        movwf   PORTB           ;this bit at least 13us · Stop bit
        nop
        nop
        nop
        nop
        nop
        nop
        nop
        movf    ADRES,0         ;we had to wait at least 80us before getting result
        movwf   LAST_RDG
        nop
        nop
        nop
        nop
        nop
        nop
        nop
        goto    loop            ;do this forever  (99 instructions (100 cycles))
END
```

1. Trigger the A/D conversion process
2. Send the previous data from LAST_RDG, using the time taken for the serial byte to be sent for the A/D to complete its conversion.
3. Read the A/D converter
4. Write the result to LAST_RDG.
5. Repeat loop continuously

The PC software has to either read the data at the rate it is being supplied (10000 samples per second) or in bursts by enabling and disabling the serial interface as required so that data arriving at the serial port is discarded. The accurately defined 10kHz sampling rate means this interface is suitable for digital signal processing type functions on the PC which are covered in more detail in Chapter 8.

An enhanced version of this interface can be used in conjunction with some dedicated PC software for audio signal measurements.

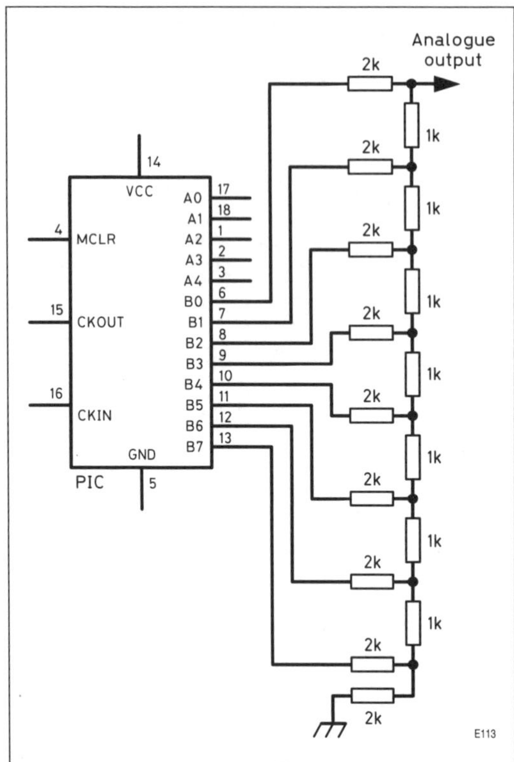


**Fig 4.5. External R/2R ladder for low-cost D/A converter**

# D/A conversion

If an analogue output is needed from the PIC software then some sort of external converter is always needed as no PIC devices at present include a D/A converter.

Several routes are possible for this task, and which one is chosen depends on a number of factors, such as how many I/O pins are available, how fast the conversion has to be performed, and the required resolution or accuracy.

In order of decreasing speed of operation, the main options are:

1. *External parallel D/A converter chip.* An eight-bit device takes up eight I/O pins and conversion is usually made as quickly as the word can be written to the I/O port. A low-cost alternative is to use an external R/2R ladder made up from a discrete chain of resistors, such as that shown in Fig 4.5.

2. *External serial D/A chip.* These use two or three serial lines in the same way as serial A/D converters. They are slower than a parallel type as the data has to be clocked out from the PIC into the D/A chip, but higher resolutions are possible, at least to 16 bits.

3. *Pulse width modulation.* This requires only an external resistor/capacitor for filtering. The PIC software generates a pulse at a regular timed interval – ideally by making use of the internal timer with interrupt-driven software. The width of this pulse is made proportional to the digital value being represented, so that the average level of the output waveform then varies in proportion and is updated at the pulse repetition rate. This is the slowest of the three D/A conversion methods, but only requires one output pin and a minimum of external hardware. A generic subroutine for generating PWM is given in Table 4.6 – this routine should be called at regularly defined intervals, and the unit delay interval adjusted to produce an output waveform with a useable min/max duty cycle. For example, if the routine is called every 10ms a suitable unit delay would be 38µs as the maximum duty cycle would then equate to $255 \times 38\mu s = 9.69$ms, very close to a 0 to 100% duty cycle variation.

## Displays and I/O for microcontrollers

Some sort of manual input and output of data is very often needed in any project making use of a microcontroller. This may be as simple as reading the state of a pushbutton switch with output to a LED being turned on or

**Table 4.6. Generic subroutine for generating a pulse width modulated output**

```
PWM                        ;Enter with 8 bit value in W register
      movwf  Temp
      btfsc  STATUS, Z     ;If zero, don't do anything
      goto   GetOutPWM     ;Could use return statement here, but that is
                           ;  inelegant programming technique !
      clrf   Counter       ;Effectively loads 8 bit counter with a value of 256
      bsf    PWMLINE       ;Initially set PWM output high
PWMLoop
      decf   Temp          ;Decrement vales each pass through loop
      btfsc  STATUS, Z     ;Reset PWMLINE when loop has passed through enough
      bcf    PWMLINE       ;  times to decrement value to zero
      call   PWMDelay      ;Or use nop statement(s), adjust delay to give a useful
      decfs  Counter       ;  min / max duty cycle.
      goto   PWMLoop       ;Always 256 loops
GetOutPWM
      return
```

off, or as complex as an alphanumeric keyboard with a text-based display
module. The PIC family of microcontrollers has been designed with I/O
very much to the fore, and some hardware features are included to simplify
interface to keyboards and switches.

## Switch and keyboard input

Fig 4.6 shows two switches connected to a pair of pins on a PIC controller.
All device families have a facility on PORTB to set an internal pull-up to the
supply rail, so that no external resistor is required. This pull-up, which
consists of a current source of approximately 100µA per pin, is activated by
clearing the RBPU bit in the PIC Option register. Another useful facility is
the Port B change interrupt. By setting the Interrupt Control register
appropriately, any change from '0' to '1' or from '1' to '0' on any of Port B4–
B7 pins will generate an interrupt. In conjunc-
tion with the Wake-up on Interrupt function,
this facility is invaluable for hand-held devices
requiring ultra-low power consumption with a
long battery life. Wake-up on keypress is set up
as follows.

1. The circuitry around the keyboard interface
   is arranged to change the state of one or more
   of PortB 4–7.
2. Wake-up on Interrupt is enabled.
3. The processor is put into sleep mode where
   is consumes very little current.
4. When a key is pressed, the processor wakes
   up and reads the state of PortB to see which
   switch was activated.
5. Any action is performed and the software
   then goes back into sleep mode.

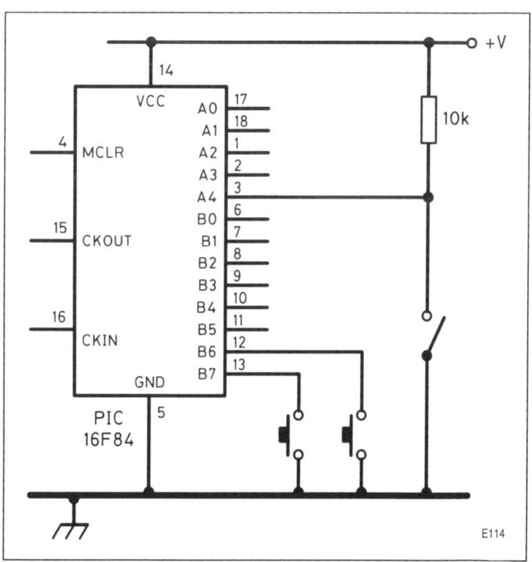


Fig 4.6. Interfacing two pushbutton switches to PortB. PortB has internal pull-up so no external resistors are required

The single switches can be replaced by a matrix-based keyboard, with the columns connected to ports B4–B7 and the rows driven by B0–B3, each pushbutton switch being connected across a row/column intersection. Before entering sleep mode, all B0–B3 outputs are set to a low state and the pull-ups are enabled. Now, if any one of the 16 possible keys in the matrix are pressed, one of the B4–B7 lines corresponding to the appropriate column is pulled low and the processor wakes up. By scanning a '1' across each of the B0–B3 in turn and reading the resultant state of B4–B7, the button that was pressed can be determined. A hand-held remote control interface using this technique is shown in Fig 7.1.

## Display output

Apart from single status LEDs or other simple types of output, alphanumeric data can be presented to users on seven-segment LED displays, or by a dedicated LCD ASCII text module. LED displays are usually operated in multiplexed fashion, where one port drives the segments and another port is used to address each digit in turn. By switching from one digit to the next rapidly enough, the display can appear to be on continuously. If the segments are driven directly from the PIC then seven pins are required, or eight if the decimal point is included. One pin is needed for each digit so it possibly to rapidly use up all the I/O pins just on the display alone. However, for driving just a couple of digits for example, no extra circuitry is needed and this is the simplest solution.

One way to save pins is to make use of a BCD to seven-segment decoder chip like the 74HC47 to drive the segments, which then only requires four lines to be allocated – all 10 digits are available as well as a few special symbols and a blank.

The driver chip concept can be extended further by using a binary decoder like the 74HC138 to decode one line at a time from a BCD digit drive. Using this technique we can now drive up to eight seven-segment displays by using just seven I/O lines, four for the segments and three for digit selection.. However, the requirement to provide decoded positive drive for one side of the matrixed display – usually the digit anodes – does mean that extra driver hardware in the form of some PNP transistors and a handful of resistors is now needed, increasing hardware complexity. The GPS clock circuit, shown later as Fig 7.7, details how a complete decoded and multiplexed display is connected. Software to drive the display has to call up the data for each digit in turn and place this on the segment drive pins. Then the code for that digit has to be placed on the address pins, the whole sequence being repeated continuously at a constant rate to avoid flickering or brightness differences. The generic code for driving a fully decoded display is listed in Table 4.7.

## LCD modules

For a more flexible alphanumeric output, modules are readily available that take ASCII codes and display them directly, with all decoding and

**Table 4.7. Software for driving multiplexed seven-segment display**

```
DisplayLoop                     ;Often part of MainLoop,
    movf   D1, W                ;Get first digit value, address 0
    andlw  b'00001111'          ;Mask to lowest bits just in case
    movwf  PORTB
    call   MpxDelay             ;Low enough to reduce flicker, say 1ms
    movf   D2, W
    andlw  b'00001111'
    iorlw  b'00010000'          ;Address of second digit
    call   MpxDelay
    movf   D3, W
    andlw  b'00001111'
    iorlw  b'00100000'
    call   MpxDelay
             ||
             ||
    movf   D8, W
    andlw  b'00001111'
    iorlw  b'01110000
    call   MpxDelay
  {place other service functions here, eg reading data, switches etc.}
    goto   DisplayLoop
```

display formatting being taken care of by a dedicated display controller chip. Modules are manufactured in a wide range of sizes from one line of 20 characters up to several lines of 80-character length. The first were those produced by Hitachi, but many different manufacturers have since adopted the concept, with several having retained the original set of commands and set-up codes – these are not elegant from a programmer's point of view but are now an industry standard.

All have the same hardware interface. An eight-bit or four-bit wide data bus is used to enter the character data – the width is defined in the first set-up codes to be sent and some crafty coding is needed to allow these to be decoded properly before the width is defined. Other lines needed are a register select line and a data strobe. A read/write line is also included, allowing data to be read from the module (the data lines can be bi-directional), but for simple display purposes this is usually tied to ground – logic '0', the write state. So, as a minimum, six wires are needed to drive these display modules in 'nibble' (four-bit) mode, and 10 for byte-wide mode. However, by suitable circuit design, the data lines can be used for other functions such as driving serial peripheral chips as these lines are redundant, and their state irrelevant, when the LCD module is not being addressed or written to. Of course, the other function or chip must be unaffected by the data lines changing as the display is addressed, usually ensured by dedicating a separate strobe or clock signal.

Table 4.8 lists the lines of code necessary to set up and initialise a typical LCD display module. Characters are subsequently sent to the display by loading their ASCII code into the W register and calling the DataByte routine. Functions such as carriage return, clear display etc are not

**Table 4.8. Software for setting up and driving Hitachi-type LCD displays**

```
;LCD Display Driving and Set up routines.
        #define     LCDRS  PORTB, 1      ;Register select
        #define     LCDRW  PORTB, 2
        #define     LCDE   PORTB, 3      ;Strobe line
        #define     LCDPORT        PORTB ;Data lines on bits 4-7, nibble (4bit) mode


============================================================

LCDInit
        movlw   d'10'               ;50ms start up delay called up by specification.
        movwf   Counter
StartDel
        call    Delay5
        decfsz  Counter
        goto    StartDel
        movlw   0x30
        movwf   LCDPORT             ;R/W - Write  RS = 0   E = Low
        nop
        bsf     LCDE                ;First send three $30's, separated by 5ms
        nop
        bcf     LCDE                ;strobe in first $30
        call    Delay5
        bsf     LCDE
        nop
        bcf     LCDE                ;second $30
        call    Delay5
        bsf     LCDE
        nop
        bcf     LCDE                ;third $30
        call    Delay5
        movlw   0x20                ;set nibble node
        movwf   LCDPORT
        nop
        bsf     LCDE
        nop
        bcf     LCDE
        nop
        movlw   0x28                ;Sets 4bit, 2 lines 5x10 dots
        call    CmdByte
        movlw   0x1C                ;Set display shift, shift right
        call    CmdByte
        movlw   0x0F                ;Display on, Cursor on, Blink
        call    CmdByte
        movlw   0x04                ;Increment cursor
        call    CmdByte
        movlw   0x01                ;Clear and Home
        call    CmdByte
        call    Delay5
        return
;------------------------------------------
CmdByte                             ;Send command byte (RS =0)
        movwf   Temp
        andlw   0xF0                ;mask to high order bits
        movwf   LCDPORT
```

**Table 4.8 *(continued)***

```
        nop
        bsf     LCDE
        nop
        bcf     LCDE            ;Strobe in high order nibble
        nop
        swapf   Temp , w        ;load low order nibble
        andlw   0xF0            ;mask to high order bits
        movwf   LCDPORT
        nop
        bsf     LCDE
        nop
        bcf     LCDE            ;Strobe in low order nibble
        call    ShortDelay
        return
;------------------------------
DataByte                        ;Normal Data byte with RS = 1
        movwf   Temp
        andlw   0xF0            ;mask to high order bits
        movwf   LCDPORT
        bsf     LCDRS
        nop
        bsf     LCDE
        nop
        bcf     LCDE            ;Strobe in high order nibble
        nop
        swapf   Temp , w        ;load low order nibble
        andlw   0xF0            ;mask to high order bits
        movwf   LCDPORT
        bsf     LCDRS
        nop
        bsf     LCDE
        nop
        bcf     LCDE            ;Strobe in low order nibble
        call    ShortDelay
        return
;------------------------------
Delay5                          ;5ms delay
        movlw   d'113'
        movwf   Temp            ;Don't use DelCount here, it's needed below!
deloop
        call    ShortDelay
        decfsz  Temp
        goto    deloop
        return
;------------------------------
ShortDelay                      ;41us delay for strobe pulses etc.
        movlw   d'12'
        movwf   DelCount
deloop2
        decfsz  DelCount
        goto    deloop2
        return
;------------------------------
        end
```

implemented with their ASCII equivalents, but are generated by calling the CmdByte routine with the appropriate generic code.

## The Atmel AVR controller

The Microchip PIC is not the only microcontroller available for experimenters, although it is probably the best-known one, albeit not the fastest. For significantly higher operation speed, Atmel's AVR microcontrollers come in. These have a RISC core running single-cycle instructions and a well-defined I/O structure that limits the need for external components. Internal oscillators, timers, UART, SPI, pull-up resistors, pulse width modulation, ADC, analogue comparator and watch-dog timers are some of the features included within AVR devices. AVR instructions are tuned to decrease the size of the program, whether the code is written in C or assembly language. With on-chip, in-system programmable Flash and EEPROM, the AVR is a perfect choice in order to optimise cost and get product to the market quickly.

The AVR series of microcontrollers was designed with speed and ease of programming in mind. They use a reduced instruction set (RISC) architecture which is optimised for speed and code density. The models with SRAM have a well-developed stack structure which is specifically designed for compact code generation from C compilers.

AVRs use a Harvard architecture which, as noted earlier, means that the program memory and buses are separate from the data memory and buses. AVRs, like PICs, have a series of models with differing feature sets. The family was made popular by the AT90 series of parts. These all have the following features:

- In-system programmable 16-bit wide Flash program memory.
- 32 general-purpose registers (usable for arithmetic instructions and as memory).
- One or two counter/timers (eight-bit/16-bit) with various programmable clock sources and prescalers.
- A variety of interrupt sources including external inputs, counter/timer overflow, UART and analogue comparators.
- Watchdog timer to guard against code crashes.
- Several power-down modes for battery-saving operations.

The members of the family have the on-board features shown in Table 4.9.

The AVR architecture allows most instructions to execute within a single clock cycle. This enables the 12MHz versions to execute code at almost 12MIPS which is an astonishing speed for this class of processor.

Reasons to use the AVR include:

- RISC architecture allows very fast and powerful applications.
- Good development support with assemblers, Basic and C compilers..
- Ease of development with in-system programmable Flash program memory

**Table 4.9. Atmel AT90 family features**

| Model | Pins | FLASH program memory (16-bit words) | SRAM (bytes) | EEPROM (bytes) | Programm-able I/O pins | Timer/ counters | External interrupts | Analogue compar-ator | UART |
|---|---|---|---|---|---|---|---|---|---|
| AT90S1200 | 20 | 512 | – | 64 | 15 | 1 | 2 | yes | |
| AT90S2313 | 20 | 1k | 128 | 128 | 15 | 2 | 2 | yes | yes |
| AT90S2323 | 8 | 1k | 128 | 128 | 3/5 | 1 | 1 | | |
| AT90S2343 | 8 | 1k | 128 | 128 | 3/5 | 1 | 1 | | |
| AT90S4414 | 40 | 2k | 256 | 256 | 32 | 2 | 2 | yes | yes |
| AT90S8515 | 40 | 4k | 512 | 512 | 32 | 2 | 2 | yes | yes |

- Sophisticated reset and interrupt handling makes for advanced programming.
- High level of on-chip peripherals – especially on the 20-pin AT90S2313.
- Several levels of power management which are excellent for battery-powered applications.

An application that makes use of the high-speed capability of the AVR is the MiniDDS application designed by Jesper Hansen and described in more detail in Chapter 7. It is an excellent example of the processing speed of the AVR processor, allowing a low-frequency direct digital synthesiser to be implemented in software at a surprisingly low cost. When run at the suggested clock frequency of 11.06MHz the DDS can generate accurate sine, triangle, sawtooth and square waves at all frequencies from 0 to over 300kHz in steps of 0.073Hz.

# References


[1] Arizona Microchip website for PIC data sheets and development software: www.microchip.com.
[2] MeLabs website for EPIC programmer details and software: www.melabs.com/products/epic.htm.
[3] Wisp628 Programmer details are available from: www.voti.nl/wisp628/.
[4] Data on the AVR microcontroller can be obtained from the Atmel website: www.atmel.com/atmel/products/prod23.htm.

# CHAPTER 5

# Simple projects using the PIC

In this chapter we show some standalone projects using PIC devices. Complete software listings are not given for these designs as many of them are quite long and include subroutines and functions common to other projects in this book. The complete software can be obtained from the RSGB website. Several of the projects in this chapter were designed during the early days of their respective authors' introduction to PIC programming so the code is often non-optimum and rather inelegant by modern standards, another reason for not giving the full printed listings!

## The G0IAY programmable beacon keyer

This keyer is designed to repeatedly send a pre-programmed message at regular intervals for identifying a beacon or repeater, for testing purposes, or for calling CQ in contests etc. A range of delays may be programmed between the CW message intervals, and as well as the normal keyer line a transmit/receive switching function is made available so the unit can take over full control of a transceiver for transmit/receive tests and CQ calls. As an additional luxury, CW speed can be changed during the transmission cycle by stored commands – allowing for example a callsign to be sent at a slow speed for identification, followed by locator information etc at a higher speed.

Unlike other beacon keyer designs published over the years, this unit can be reprogrammed without having to remove the controller chip, merely by connecting it to a PC running a terminal programme such as Hyperlink over an RS232 interface. The CW message is stored in the non-volatile EEPROM storage available in the 16C84, 16F84 and later devices.

The circuit is given in Fig 5.1 – the design is not much more complicated than that of the bare minimum PIC chip itself with two transistors to interface to standard keying and transmit/receive control circuits. A PCB layout making use of surface mount components to produce a postage stamp-sized module is shown in the photograph in Fig 5.2. The listing KEYER2.ASM is an example of the inelegant first programmes we often write until programming experience grows. However, it does work and so has been included in its original form; much 'cleaner' CW generation and RS232 routines are

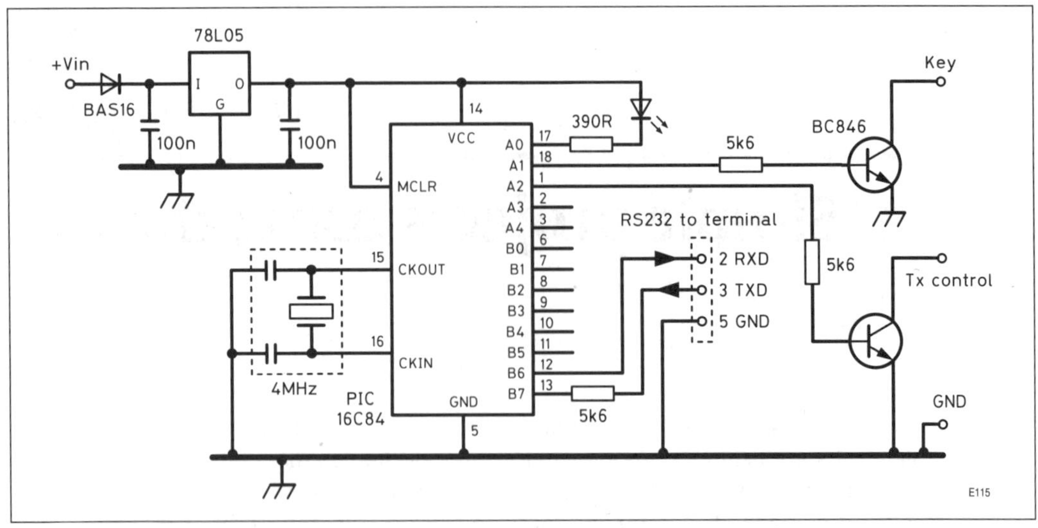

*Top:* **Fig 5.1. The G0IAY beacon keyer circuit diagram**

*Above:* **Fig 5.2. Photograph of the keyer module**

described elsewhere in this book. The commands needed to programme the keyer message over the RS232 interface are shown in Table 5.1.

## SWR meter display

In this project, the two voltage levels generated by a SWR meter head, corresponding to forward and reverse power, are measured by the A/D converter inside a PIC 16C71, then used to generate an automatic, power-independent, SWR reading on an analogue meter. No adjustment is needed as power level is changed as the correction for forward power reading is calculated automatically. In addition, a bar graph is presented showing actual forward power. There are plenty of suitable RF head designs for SWR measurement on all the amateur bands available so none will be reiterated here. Whatever RF head is adopted, it should be designed to give a voltage output of no more that 5V when the maximum power is being measured. An equal level for forward and reverse voltages is assumed, ie the head unit is properly symmetrical. The unit can be battery operated and provision for automatic turn-on is included, so the PIC and display circuitry are powered up by detecting a level on Vfwd when a few watts of RF are applied. The unit automatically turns off after three minutes with no RF. No circuit details are given as constructors will probably want to choose their own way of implementing these functions.

The 16C71 has up to four analogue inputs available – two of them are used here with the 5V Vcc supply being used as the A/D reference voltage. The main loop of the programme measures Vfwd and Vrev to eight-bit accuracy, then calculates the ratio Vrev/Vfwd which is needed to give a

**Table 5.1. Programming instructions for the IAY Beacon Keyer module**

Set the terminal programme (Hyperlink, Procomm etc) to 1200 baud, no parity, 8 bits, 1 stop bit, no handshaking. If the interface is connected when power is first applied, the keyer will go into command mode, where messages can be displayed and entered, and WPM and keying polarity can be set. Valid commands are:

V   Display software version.

S   Go into beacon mode (if this command is executed, then the keyer will need to be powered down to re-enter command mode).

F   Display keying polarity and WPM flags.

L   Set keying polarity and WPM flags.

Notes: Keying polarity can be − or +
− denotes transistor on = keyed (ie key line grounded = keyed)
+ denotes transistor off = keyed (ie key line floating = keyed)

WPM is set by typing in a letter:

A = 5WPM       B = 8WPM       C = 10WPM       D = 12WPM
E = 15WPM      F = 18WPM      G = 21WPM       H = 25WPM

D   Display message in EEPROM.

E   Enter new message.

Notes: Just type in the message. Procedural characters are coded as follows:
[ − CT
] − VA

There are several commands that can be embedded into the message. They are all entered in the form <command> and each use one character of message storage. Messages can be up to 63 characters long.

<R>    − Return to start of message with no delay.
<Wn>   − Set WPM to n (n is a letter, as listed above)

This command overrules whatever is set in the flags, so if keying speed is changed during a message, it will remain at that value until changed back.

<Dxyz> − Delay mode.
x denotes transmit (T) or receive (R)
y denotes key down (D) or key up (U)
z is delay:
A = 15 seconds     B = 30 seconds     C = 45 seconds
D = 1 minute       E = 1.5 minutes    F = 2 minutes
G = 4 minutes

Examples:
<DTDF> - Transmit with key down for 2 minutes.
<DRUB> - Receive with key up for 30 seconds.

If the keyer has been connected to allow control of transmit, it can be programmed to do automatic CQ calls with periods of receive in between calls.

It is advisable to always put an <R> at the end of the message, as the software does not do this automatically. The only exception to this is if the message is 63 characters long, where the message will automatically repeat. Message input can be ended by pressing Return.

A typical message would be:

GB3SCX <WE>GB3SCX IO80UU59 <DTDA<R>

This sends the beacon callsign once at 10 WPM, then again at 15 WPM followed by the locator, also at the faster speed. Then follows a delay of 15 seconds of plain carrier in transmit mode with the key down, after which the sequence repeats.

**Table 5.2. Division routine used within the automatic SWR bridge software**

```
        clrf    VREVLO              ;Promote Vrev to 16 bit and divide by Vfwd
        movlw   0xFF                ;always at least one incr so start with -1
        movwf   SWR                 ;also sets default for overflow as $FF
        movf    VFWD,w              ;
        subwf   VREV,w              ;First check for VREV - VFWD (= W)
        btfsc   STATUS , C          ;if positive (C=1) then overflow will occur
        goto    GetOut              ;leave SWR at FF to give maximum
        btfsc   STATUS , Z          ;same applies for zero (equal)
        goto    GetOut

DivLoop                             ;
        incf    SWR                 ;First pass sets it to zero if FF
        movf    VFWD,w
        subwf   VREVLO              ;VREVLO = VREVLO - VFWD
        btfsc   STATUS , C          ;C = 0 if borrowed, ie -ve
        goto    DivLoop
        movlw   1
        subwf   VREV                ;can't use decf as we need C bit set
        btfsc   STATUS , C
        goto    DivLoop             ;if clear, now -ve so finished division
GetOut
```

power-level-independent measurement of the SWR. Division is one of the more complex mathematical routines to implement on any digital platform, and here a crude division routine operates as follows.

First Vrev is compared with Vfwd to see if it is higher – it should never be the case, but an unbalanced RF head may give the erroneous result at high SWR levels; SWR is defined as maximum in this case and no calculation performed. The case of zero Vrev, corresponding to 1:1 SWR, is also trapped to prevent a divide-by-zero problem. If the two voltage measurements give valid numbers, the eight-bit value for Vrev is expanded to 16 bits by appending a low-order byte consisting of all zeros, then the eight-bit value for Vfwd is repeatedly subtracted from this 16-bit number until the result of the subtraction is negative. Each subtraction increments a counter (the SWR variable) so that when the result goes negative, this value contains the eight-bit result of the integer division Vrev/Vfwd. Table 5.2 gives a listing of the code included within the division routine.

The output to drive an analogue meter to show the SWR is generated by pulse width modulation. Inside a loop the SWR variable generated from the division process is decremented by one for each pass. The output pin that drives the analogue meter is initially set to a '1', but as soon as the SWR variable reaches zero this pin is set to a '0'. As the loop is always 256 clocks long the mean voltage on this pin after passing through a CR filter is proportional to the SWR value. A delay is built into the loop so that the SWR display is updated around 10 times per second. If it was set much faster than this, the extra unknown time taken for the division process – anything from 0 to over 2000 clock cycles – could affect the PWM linearity. As it is, maximum error introduced gives around 10% longer cycle time at maximum SWR and the direction of this error is such as to slightly compress the top end of the scale – not usually a problem as the SWR is usually too high to be useful at this point.

Next, the forward voltage reading is used to drive the bargraph display on PORTB. A pseudo-logarithmic scale is employed here to make the bargraph display more useful and is generated by taking the three most significant bits of the reading only, then lighting, successively, from zero to eight LEDs in the bargraph depending on the value in these MSBs.

The complete software listing is entitled 'SWRBR3.ASM' and can be obtained from the RSGB website.

## Frequency measurement

The 16F84, along with most other PICs, includes at least one counter timer within the range of peripheral functions, usually with an associated prescaler which divides the input signal by one of several programmable values. This can be driven from either the internal clock oscillator, or from an external signal connected to one of the pins designated for this function; on the 16F84 this is Pin 3, corresponding to input A4. The prescaler is designed to be clocked at a much higher frequency than that at which the rest of the circuitry operates, with operation at 30–40MHz possible. A frequency counter function can be generated by enabling and disabling the counter in a controlled way using timing derived from the processor clock while it is counting this external signal.

Apart from displaying the frequency as in a conventional frequency meter, simple straightforward frequency measurement opens up all sorts of possibilities for add-on accessories and automation in the areas of HF transmitting, such as automatic antenna selection based on measured frequency, and selection of filters and PA stages.

The simplest way of using the counter is illustrated in the part listing shown in Table 5.3 which gives the set-up data needed to operate the counter timer with an external input and a subroutine which generates a value in the variable W register that is proportional to the frequency being measured. The resolution of the counter depends on the gate timing interval, defined by the loop around the label `GateLoop`, the value of the constant `GateConst`, the processor clock value, and the prescaler setting.

In the listing shown, the prescaler is set to divide by eight, the delay loop generates a delay of $5.N + 5$ clock cycles so, with a value of 99 for the constant and a 1MHz clock from a 4MHz crystal, the gating, or timing, interval is 500µs, measuring the frequency input divided by 16. The resulting frequency is therefore measured to a resolution of 16kHz with a maximum upper frequency limit of 4.08kHz dictated by the eight-bit count limit on the timer.

The counter resolution can be extended by making use of the counter overflow bit 2 in the INTCON register. Every time the counter overflows from 255 back to a count of 0, this bit is set. An interrupt can be generated, or the bit can be read by polling, then, by incrementing a software counter

**Table 5.3. Setting up and using the counter/timer function for frequency measurement**

```
;Setup needed to initialise timer.
     bsf      STATUS,RP0       ;ram page 1
     movlw    b'11100010' ;External input, Prescalar /8
     movwf    OPTION_REG
     movlw    b'00010000' ;
     movwf    TRISA        ;A4 as input for Fin, rest as outputs
     bcf      STATUS,RP0        ;ram page 0


     GateConst     equ    d'99' ;Depends on clock frequency, prescale etc.
----------
;Frequency measurement routine
GetFreq
     clrf     TMR0         ;Initialise the timer
     movlw    GateConst          ;This constant determines how long the count
     movwf    DelReg             ;  gate is opened for.
     nop
     nop
     nop
GateLoop              ;delay between TMR set and read = 5N+5, so a value
     nop                 ; of d'99' gives 500us assuming a 4MHz xtal.  with
     nop                 ; /8 prescale, Temp contains multiples of 16kHz
     decfsz   DelReg
     goto     GateLoop
     movf     TMR0 , w           ;Read the counter at precise timed instant
     return               ; Frequency measurement returned in W
```

each time, a frequency counter of arbitrary length can be generated. With a suitable gating time, higher-resolution measurement is possible.

Using the prescaler in this way limits the measurement accuracy possible as the inherent accuracy of ± one count applies to the divided-down waveform. Therefore for high-division ratios, accuracy is limited to a count of ± prescaler division ratio.

Peter Rhodes, G3XJP, has written a number of programmes for the PIC that use a more precise method of frequency measurement, where optimum resolution is gained by clocking out the residual count left in the prescaler after each increment of the counter register [1, 2].

## Waveform generation

Sometimes a PIC can be a simple solution for an application where the use of a processor would not normally be considered. An example of this is waveform generation. By the addition of a D/A converter, a sampled waveform of any arbitrary shape can be formed by generating, or looking up, successive values and sending these to the D/A converter. A normal D/A converter may not even be necessary as illustrated in the sine-wave generator circuit of Fig 5.3. This particular function simulates a sine wave by generating the software equivalent of a shift register in a loop, with feedback from the final stage to the first stage passing though an inverter. The concept is covered in more detail in reference [3].

## Synthesising waveforms

The weighted resistors are chosen so that, as the '1's or '0's progressively build up from left to right, a stepped approximation to a sine wave is generated. With an eight-bit shift register clocked at 16 times the wanted frequency output, a 16-step approximation to the required waveform results, and theory predicts that, apart from the fundamental, the only spurious products will be at $15.F$, $17.F$, $31.F$, $33.F$ and so on, at levels of 1/15, 1/17, 1/31 of that of the fundamental. A simple filter is all that is then needed to clean up the output, resulting in a stable and clean sine wave. Other waveforms can be generated by choosing appropriate resistor values to give the required increase/decrease in amplitude for each step.

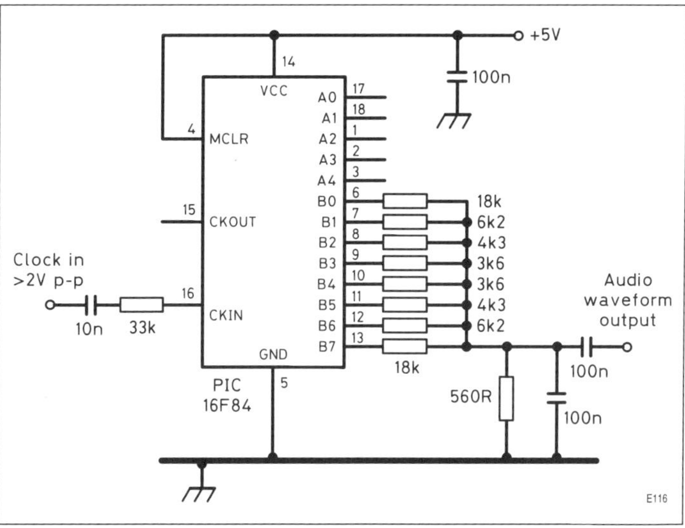

**Fig 5.3. Sine-wave generator circuit diagram**

The PIC source code shown in Table 5.4, when used with the circuit of Fig 5.3, takes this waveform generator a stage further, producing a binary phase shift modulated carrier of 1.6kHz, modulated with a 255-bit pseudo-random sequence at 400 bits/second. While of limited use as it stands (the pseudo-noise waveform was used to evaluate a microwave data communications link), the listing illustrates how modulated waveforms can be easily generated.

## Interrupt-generated timing

Here a timer-generated interrupt is used to define the precise timing interval, relieving the programmer from the tedious job of carefully adjusting clock cycle delays to achieve the correct result. Instead, the timer counter module generates a count which is derived by dividing down the processor clock by a prescaling factor, here four times. Every time the timer/counter overflows, an interrupt is generated which sends the software to the interrupt vector at programme address 004 – given the label `ints` in Table 5.4.

The following method is used to adjust the interrupt frequency to the wanted values. At the start of the interrupt code the timer is preloaded with a value of 233 so that it only has to count from this value up to 256 before overflowing, a count of 23. The number is calculated by noting that the interrupt always takes several clock cycles to process (here seven clocks) so by the addition of an extra `NOP` statement the overhead consumes eight clock cycles. As the timer operates on the clock frequency divided by four, these eight cycles equate to two counts of the timer, hence the count of 23 before the interrupt is generated gives an effective division by 25. The interrupt frequency is calculated by dividing the processor clock frequency

**Table 5.4. Complete listing of BPSK pseudo-random sequence generator**

```
;PRNGEN02.ASM
;Generates 1.6kHz sinewave, BPSK modulated with PR sequence.  10.24MHz Xtal,
;2.56MHz clock, prescale /4, timer counts 25 cycles for 25.6kHz interrupt


        LIST  P=16F84


        INCLUDE "P16F84.INC"
        cblock 0x0D
            Counter1              ;Times data clock from interrupt
            SineData             ;Makes up synth sinewave, BPSKed for PORTB
            SR1                  ;Shift register
            SR2
            Temp
            FB                   ;Contains Feedback for PN generator
        endc


        #define   Tap1    SR1 , 0    ;Feedback taps for PR sequence
        #define   Tap2    SR1 , 1
        #define   Tap3    SR1 , 3
        #define   Tap4    SR1 , 5


        __config    0x3FF1

;code starts here


        org     0
        nop
        movlw   b'00000000'      ;No interupts until passed thro here once
        movwf   INTCON           ;
        goto    startup          ;jump to main code
;————————————————————
ints                             ;Here 4 cycles after timer overflow
        bcf     INTCON , T0IF    ;  every 100 clocks = 25.6kHz sampling rate
        nop                      ;Needed to make up 8 clock cycles total
        movlw   d'233'           ;=N :  Timer counts 25 of (Fc / 4)
        movwf   TMR0             ;Timer restarts counting 2 cycles after here.
                                 ;To here takes 8 clocks (with the single NOP)
                                 ;N = 256 - Count + 8/P
 ;Below here not time critical, but must be less than than 95 cycles total


        movf    SineData , W     ;Output the sample first, so there is no
        movwf   PORTB            ;  jitter on the waveform


    ;Generate 1.6kHz sinewave using 8 bit reg with weighted resistors


        rrf     SineData , W     ;LSB into carry, but preserve SineData
        rlf     Temp             ;Carry into Temp,0
        comf    Temp             ;Toggle LSB = FB
        rrf     Temp             ;FB into Carry
        rrf     SineData         ;do the shift, Feedback into MSB


    ;Test for Clock change
        incf    Counter1         ;Count 64 = 0x40   for 400b/s
        movlw   0x40
```

**Table 5.4 (continued)**

```
        subwf   Counter1 , W    ;Test for count
        btfss   STATUS , Z
        goto    NoClock
        clrf    Counter1        ;13 clocks to here

        movlw   0               ;
        btfsc   Tap1
        xorlw   1               ;Toggle W=0/1 if FB bit is a 1
        btfsc   Tap2
        xorlw   1
        btfsc   Tap3
        xorlw   1
        btfsc   Tap4
        xorlw   1               ;W contains data to be shifted into SR MSB
        movwf   Temp            ;Put it into Temp,0
        rrf     Temp            ;and into C
        rrf     SR1             ;then do the shift, with C into MSB of SR

        btfsc   SR1 , 0         ;Use LSB of SR as data
        comf    SineData        ;Differential modulation, change if -ve
NoClock
        retfie
;--------------------------
startup
        clrwdt
        bsf     STATUS,RP0      ;ram page 1
        movlw   b'11000001'     ;Internal input, Prescalar /4 no pull-ups
        movwf   OPTION_REG
        movlw   b'00000000'     ;
        movwf   PORTA           ;
        movlw   b'00000000'     ;All outputs for testing
        movwf   PORTB
        bcf     STATUS,RP0      ;ram page 0
        movlw   b'10100000'     ;Generate overflow on TMR0 Overflow (T0IE)
        movwf   INTCON          ;
        clrf    PORTB
        clrf    PORTA
        clrf    Counter1
        clrf    SineData
        movlw   0xFF
        movwf   SR1
        movwf   SR2

SitAndWait
        goto    SitAndWait      ;Sit in this loop while waiting for interrupt
;--------------------------
        end
```

by the prescale factor and this interrupt-generated value, ie 100. With a crystal of 10.24MHz, the processor runs at a quarter of this, ie 2.56MHz, so the interrupt occurs at a frequency of 25.6kHz.

Most of the active code is within the interrupt service, with the main loop just a place for the processor to sit and wait for the next interrupt interval.

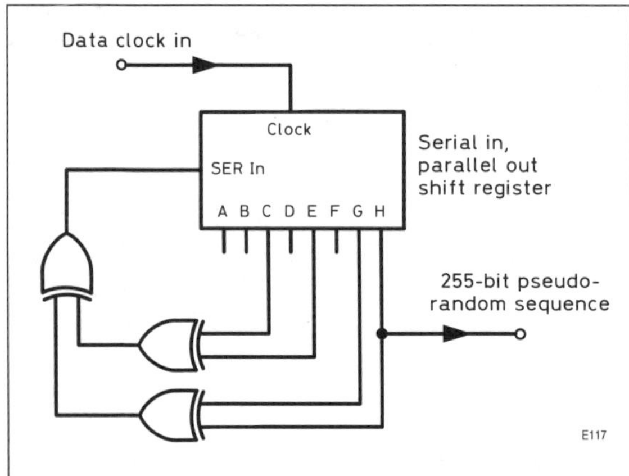


**Fig 5.4. Psuedo-random sequence generator**

Within the interrupt routine a simulated shift register with feedback, using the variable `SineData`, first generates the audio frequency of 1.6kHz , whose final frequency is equal to the interrupt frequency divided by 16. The next part of the code uses the `Counter1` variable to count every 64 interrupts, giving a data clock frequency of 400Hz (25.6kHz/64) which then passes though to the pseudo-random sequence generator routine. The 255-bit long PRS is generated by simulating the classic shift register with feedback via exclusive-OR gates connected to appropriate taps as shown in Fig 5.4. Depending on the output of the PRS at each 400Hz interval, the sine wave is inverted, resulting in a 1.6kHz tone, differentially phase shift keyed at 400 bits/second.

As the interrupt occurs every 100 clock cycles and there are always several 'hidden' clock cycles required to process the interrupt, the service routine must be short enough to complete before the next interrupt would be generated. If the processor has not seen a Return from Interrupt (`RETFIE`) command at that point the new interrupt will be delayed until it has returned, upsetting the timing interval. Comments in the listing show where code within the interrupt is time critical, and vigilance has to be constantly maintained when writing such code when using interrupt-generated timing so that the time taken to service the interrupt is always less than that of the interrupt frequency.

Having said that, the timer-generated interrupt is one of the easiest ways to produce accurate timing on a PIC without the complexity of defining exact clock cycle counts.

## Data logging

In the last chapter we looked at an A/D converter with a serial interface used for sampling an audio waveform. For monitoring of voltages that change slowly, over minutes or days, this is not the best solution to follow as the PC logging software has to cope with the huge flow of data coming in. Most of this has to be rejected and custom logging software is going to be required. So why not make use of the PIC to generate a suitable logging interval, then format the data sent along the RS232 interface?

We can then take the opportunity to make the data format more palatable to standard utility software that can directly read from the serial interface and save to disc for subsequent analysis.

The interrupt-generated timing technique in the last section can be extended, by suitable choice of counter length, to generate timing intervals from seconds to years. Additionally, the binary data as generated by the A/D converter can be converted to standard ASCII text so that it can be directly read by the PC and saved directly to disc or displayed on a screen. In either case Hypertrm or Procomm, the ubiquitous serial port software, can be used to read the incoming formatted data directly.

As a final twist, why send the data on the RS232 interface at all? Large memory chips are available these days, and many have serial interfaces that can be connected to the PIC with two or three wires. In the case of a logger that may only be called upon to generate a few kilobytes of data, the latter could all be dumped into the local memory, and this could be subsequently uploaded to a PC.

With CMOS technology throughout, the entire data logger need only consume an average of a few microamps (especially if the processor is clocked at 32kHz) so it could be left monitoring data in a remote location for many months running off a single battery. Component supplier's catalogues list many examples of suitable memory devices.

The extra routines needed to add some of these functions to the serial D/A interface of Fig 4.5 are given in Table 5.5. A single-chip, multi-purpose data logger, the HF-08, is based around the 16C710 PIC, a lower-cost alternative to the 16C71, and allows a number of sampling functions to be selected by connecting links on the module or by software commands. Sampling and data formats possible are:

- Eight-bit binary data sampled at 100µs, 140µs and 250µs.
- I/Q sampling of a 1kHz tone at 1000 samples per second.
- Sampling at 0.2, 1, 5, 10, 60 and 300s intervals with eight-bit binary output.
- Sampling as above but with ASCII text output for direct computer logging.

The circuit for the HF-08 data logger is shown in Fig 5.5, and a compact layout designed to fit inside a dual nine-way D-type connector adapter package as shown in Fig 5.6. Software for the HF-08 can be obtained from the RSGB website as LOGGER.ASM.

# The HF-08 data logger

This section is taken from the HF-08 operating manual.

The HF-08 data logger is a small eight-bit analogue-to-digital converter module for digitising an analogue signal and sending the data to the serial (COM) port of a personal computer or similar.

There is sufficient flexibility built into the design of the module to allow numerous different types of data logging when used with the appropriate software. The analogue input can be link selected for a range of input voltages and, with the addition of one or two external components, this can be extended still further.

The data output from the module is sent in RS232 format at a fixed rate of 115200 baud (bits/second). Data format is eight-bit data, one stop bit, no parity. The module is self-powered from the RS232 port using the additional handshaking lines RTS and DTR. Voltage levels on the data output

**Table 5.5. Routines needed to add more useful data logging functions to the serial port A/D converter of Fig 4.5**

```
;Set up timer interrupt


    bsf     STATUS , RP0
    movlw   b'00000100'         ;Internal input, Prescalar /32 enable pull-ups
    movwf   OPTION_REG
    bcf     STATUS , RP0
    movlw   b'10100000'
    movwf   INTCON              ;Enable Timer Interrupt, disable PORTB change
    clrf    Counter1
    clrf    Counter2


;=================================================
;Interrupt service routine to read A/D converter at required time intervals.  An 8ms
; timer overflow interrupt is set using internal 1MHz clock with /32 prescalar
; and preseting for a count of 250 before overflow.


    btfsc   INTCON , T0IF   ;test for timer overflow
    goto    ClockTick       ;Can only happen in slow modes
    retfie                  ;default for any other interrupt

ClockTick
    nop
    nop         ;Here 13 cycles after timer overflow   NOW IN INTERRUPT SERVICE
    movlw   2
    call    Delay           ;(15) need 32 cycles INT > MOVWF TMR0
    nop

    bcf     INTCON , T0IF   ;every 8ms
    nop                     ;
    movlw   7           ;Timer counts 249 of (Fc / 32), + 32 residual clocks above
    movwf   TMR0            ;Timer restarts counting 2 cycles after here

 ;--- Below here is not time critical but must complete in 8ms

    incf    Counter1        ;Count 8ms units for wanted delay
    btfsc   STATUS , Z
    incf    Counter2        ;16 bit count = max 524 seconds

    movf    TickCount2 , W      ;Value of TickCount1/2 determines total delay
    subwf   Counter2 , W    ;Test for MS Count
    btfss   STATUS , Z
    goto    DoneTick        ;Exit if not MSB count limit
    movf    TickCount1 , W
    subwf   Counter1 , W    ;Now test for LS count
    btfss   STATUS , Z
    goto    DoneTick        ;exit if not LSB count limit

    clrf    Counter1        ;Here every timed interval
    clrf    Counter2        ;Reset counter
    bsf     ADCON0,GO       ;Hit A/D converter
    movlw   d'19
    call    Delay           ;Wait 100us for conversion to complete
    movf    ADRES , W   ; A/D reading in W ready for further processing
```

**Table 5.5 (continued)**

```
    retfie

;===============================================================================
;Binary to Text conversion

SendText      ;Convert 8 bit binary value to ASCII text in range 000 to 255
          movwf   Temp          ;Save binary value
          clrf    S3            ;Hundreds
          clrf    S2            ;Tens
          clrf    S1            ;Units
HunLoop
          incf    S3            ;1 to 10 counts
          movlw   d'100'
          subwf   Temp          ;Temp = Temp - 100
          btfsc   STATUS , C    ;1 if +ve after subtraction
          goto    HunLoop       ;Repeatedly subtract 100 until -ve
          decf    S3            ;adjust to 0-9
          movlw   d'48'         ;Adjust to ASCII
          addwf   S3
          movlw   d'100'
          addwf   Temp          ;restore final decrememt so Temp is in range 0-99
TenLoop
          incf    S2            ;1 to 10 counts
          movlw   d'10'
          subwf   Temp
          btfsc   STATUS , C    ;1 if +ve after subtraction
          goto    TenLoop       ;Repeatedly subtract 10 until -ve
          decf    S2            ;adjust to 0-9
          movlw   d'48'         ;Adjust to ASCII
          addwf   S2

          movlw   d'10'
          addwf   Temp          ;restore final decrememt so Temp is in range 0-9

          movf    Temp , W
          movwf   S1
          movlw   d'48'         ;Adjust to ASCII
          addwf   S1

          movf    S3 , W
          call    Send232
          movlw   d'20'
          call    Delay

          movf    S2 , W
          call    Send232
          movlw   d'20'
          call    Delay

          movf    S1 , W
          call    Send232
          movlw   d'20'
          call    Delay
```

**Table 5.5** *(continued)*

```
        movlw   d'13'         ;Final carriage return
        call    Send232

        retfie


;-----------------------------------------------------
Delay           ;general purpose delay routine needed in int service and other places
     movwf  DelCount          ;Enter with N in W register
deloop
     nop                      ;Total delay = 5.N + 3 (+ 2 to call = 5.N + 5)
     nop
     decfsz DelCount
     goto   deloop
     nop
     return
;-----------------------------------------------------
```

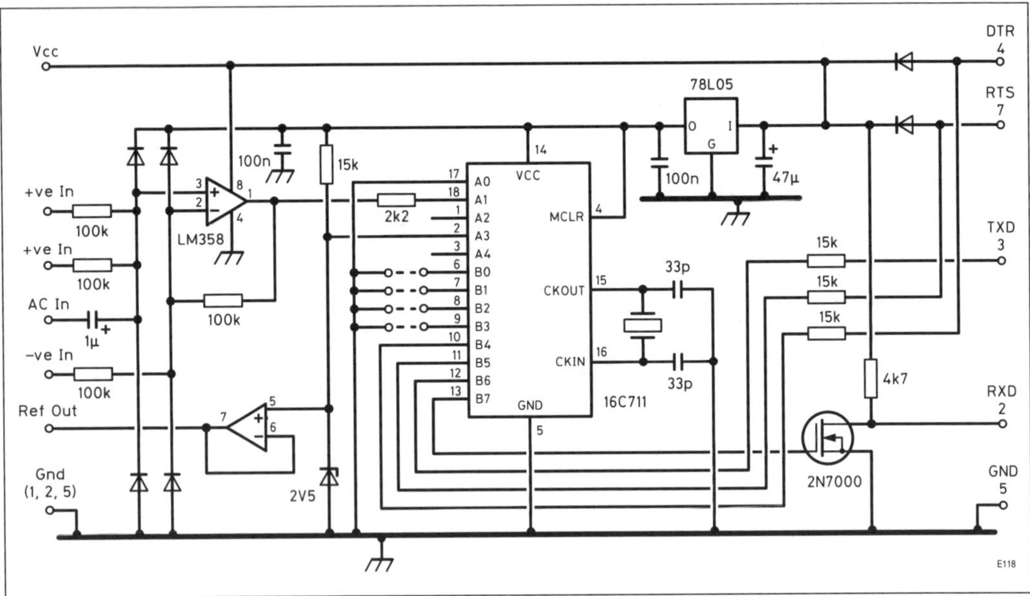



**Fig 5.5. HF-08 data logger circuit diagram**

do not follow the full RS232 convention of plus and minus levels; instead, the voltage swing is from 0 to +(5–10)V. This is permissible, as all modern RS232 receiver chips actually detect at a threshold of around 1V positive, allowing 'unofficial' compatibility with other 0–5V signalling standards.

## Connections

The serial data output comes from a nine-way D-type female connector, intended for direct attachment to a PC serial port. A short length of cable up to around 2m may be used here but the driver circuitry is not designed for runs longer than this.

## Setting the analogue input range

The analogue input is applied to the nine-way D-type male connector. Connection to various points in the input buffer amplifier are available on this connector plus a buff-ered version of the internal 2.5V reference.

By linking pins appropri-ately, the input measurement ranges shown in Table 5.6 are available.

These may be further ex-tended by the use of external resistors. Refer to Fig 5.5 for the input buffer circuit.

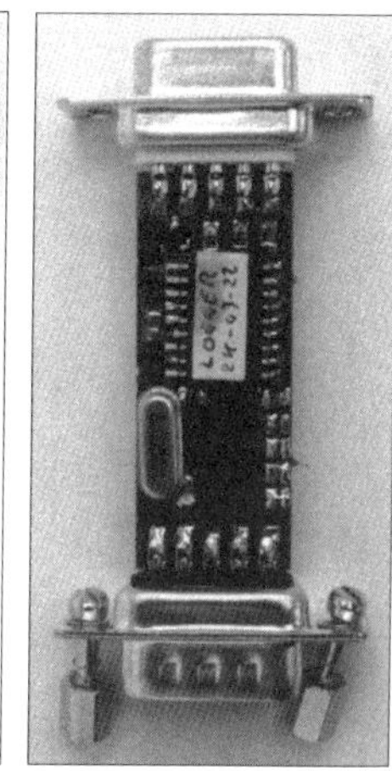

Fig 5.6. The HF-08 data logger

## Sampling modes

Three types of operation are available with the HF-08:

1. Fast sampling for audio or rapidly changing signals. Three sampling rates are available :

    | | |
    |---|---|
    | 10000Hz | 100µs sampling interval |
    | 7142Hz | 140µs sampling interval |
    | 2500Hz | 400µs sampling interval |

    Output data format is limited to one eight-bit binary data word per sam-ple. Software commands, eg to change rate, may be sent in real time while in this mode.

2. Complex sampling for vectorscope or similar purposes. A signal of 750–1250Hz, centred on 1000Hz, is down-converted to zero frequency and complex in-phase and quadrature components are outputted. Software commands will be accepted while in this mode.

3. Slow sampling modes for longer-term data monitoring. Six sampling rates are available :

| Table 5.6 | | | | | |
|---|---|---|---|---|---|
| **Range** | **Input signal (hi)** | **Input Signal (lo)** | **Link** | **Link** | **Zin** |
| 0 to 1.25V | 8 | 2 (0V) | 9 to 5 | | Very high |
| 0 to 2.5V | 8 | 2 (0V) | | | Very high |
| 0 to 2.5V (differential) | 8 | 9 | 7 to 1 | | 200kΩ (pin 8) 100kΩ (pin 9) |
| 0 to 5V | 8 | 2 (0V) | 7 to 1 | | 200kΩ |
| –2.5V to 2.5V | 8 | 2 (0V) | 7 to 3 | | 200kΩ (to Vref) |
| –2.5V to 0V | 8 | 2 (0V) | 9 to 5 | 7 to 3 | 200kΩ (to Vref) |
| ± 2.5V AC | 6 | 2 (0V) | 7 to 3 | | 100kΩ in series with 1µF |

0.2s interval
1s
5s
10s
1min
5min

Data is transmitted in real time immediately after the measurement sample is taken. Output format can be selected either as an eight-bit binary word, same as the fast sampling rates, or BCD text format consisting of three ASCII digits for hundreds, tens and units of the binary value from 0–255, followed by a carriage return (ASCII 13). This latter option is designed for direct logging to a file using, for example, a terminal programme such as HyperTrm or Procomm.

No commands sent on the RS232 link will be accepted while slow sampling is underway.

## Control of operating mode

The operating mode of the device may be set by two means:

*Software control*

When in fast sampling mode, the ASCII command Sn will software select a sampling rate from Table 5.7.

Other ASCII commands accepted while in fast sampling mode:

T     Set text mode for slow rates
B     Set binary mode for slow rates
L     Lockout (inhibit any more serial commands)

All software commands consist of just one or two characters; no carriage return is necessary. Upper case should be used for all commands. Once a slow mode has been selected, no further software commands will be accepted, so the T or B command needs to be sent before the Sn one when selecting one of these.

For example, T S 6 will set sampling at every five seconds with output data sent in text format (number 0–255 followed by carriage return). S 2 L will set 2500Hz sampling and lock out any further serial commands.

*Link selection*

As an alternative, up to four wire links may be added to set the operating mode that is selected when power is applied. In Table 5.8, a '1' means a link is fitted, a '0' indicates the link is absent.

Serial commands will be accepted while any of the first four sampling rates are selected.

## Operation

The unit is connected to a spare COM port. If any of the supplied software is used for data logging follow the start-up instructions for the appropriate programme for selecting COM port, sampling rate etc.

At start-up, the links are read and data output starts immediately at the appropriate sampling rate.

For logging of text mode results a serial terminal programme such as Hypertrm (supplied with Windows) or earlier DOS-based software such as Procomm may be used. The following procedure makes operation of the module straightforward.

Set up the terminal programme for the appropriate port, 115200 baud, no parity, eight bits and one stop bit, ie 115200 N 8 1. All flow control must be OFF. Incoming <CR> translation to <CR LF> should be set to ON. All these settings can be found in the Hypertrm Properties menu; other serial driver programmes will provide similar functions.

Hypertrm allows these settings to be written to a start-up name and an appropriate icon may be selected which makes subsequent start-up very straightforward.

If no links are fitted, the screen will fill with meaningless characters – these are the binary results of the input data being sampled at 10000 samples per second. Key in (for instance) 'TS5' and the display will change to three-digit numbers being produced every second at one per line.

If writing your own software, the following points need to be borne in mind.

Module power is obtained from the RTS and DTR lines and at least one of these, preferably both, needs to be active (high). Flow control is not available, but if necessary data can be started and stopped by powering down the module by removing RTS and DTR. Mode configuration commands will need to be resent each time the module is powered up unless link selection is in use.

Precautions need to be taken against serial buffer overflow situations arising, as data starts to be transmitted immediately after powering up.

Some laptop computers may not provide a 'strong' enough output from the RTS and DTR port to allow the module to generate the necessary regulated voltage internally. In this event, a separate battery of anything between 7 to 20V may be connected between pin 4 (positive) and pins 1, 2, 5 (ground). Current consumption is approximately 6mA.

All analogue input pins are protected against reasonable overload: internal voltage swing is clipped to the internal supply voltage and current is limited by $100k\Omega$ resistors; maximum safe overload voltage is limited by power dissipation in these resistors to 100V DC or RMS. The 2.5V reference output on pin 3 comes from an internal op-amp buffer and is protected against a short-circuit to ground by the inherent current limit provided by the power available from the COM port. Protection against connection of this to any external voltage sources is not provided.

To maximise operating speed, no decoupling of the analogue inputs has been provided. In noisy environments, or when sampling slowly changing signals, consideration should be given to the addition of decoupling or filtering capacitors.

When used with fast-changing signals, for example audio, no anti-aliasing filter is provided, and signals greater than one half the sampling rate will give erroneous results as products will be aliased. Sampling rates 1–3 have been selected as follows:

- 7142Hz (1/140µs exactly) allows the output of an SSB receiver limited to 300–3300 Hz to be optimally sampled.
- 2500Hz allows a communications receiver using CW mode with a narrow filter and BFO tone up to 90 Hz to be optimally sampled.
- 1000Hz I/Q sampling allows the special case of the phase of a tone centred on 1kHz to be detected directly. Inside the HF08, the incoming signal is effectively down-converted, or mixed, with sin and cos components of a local oscillator at 1kHz. The lack of anti-aliasing filtering and the method of conversion means that the input signal must be pre-filtered to the range 750–1250Hz. This is conveniently done by using a CW filter in a communications receiver and setting the BFO note to 1kHz or tuning for a 1kHz output frequency.

Data output is in the following form, and consists of four bytes per sample containing 10-bit I/Q data in two's complement format:

- Byte 1, Ihi, two high-order bits of the I channel (cosine) data in the form 000000xx.
- Byte 2, Ilo, lowest eight bits of I channel data.
- Byte 3, Qhi, two high-order bits of the Q channel (sine) data in the form 110000xx.
- Byte 4, Qlo, lowest eight bits of Q channel data.

The two high-order bits of Ihi and Qhi are coded as shown to permit correct identification and synchronisation should data slippage occur. Ideally, every four-byte frame of data will be checked for correct alignment before use is made of it.

Full I/Q data can be reconstructed (using 'C' notation) from:

$$I = ((Ihi \ \& \ 0x03) << 6)/64 + Ilo$$
$$Q = ((Qhi \ \& \ 0x03) << 6)/64 + Qlo$$

This will correctly align and sign extend the data to give 16-bit signed integers.

$$Phase = ATAN2 \ (I/Q)$$
$$Mag = SQRT \ (I\char`^2 + Q\char`^2)$$

## References

[1] 'The PicATUne automatic antenna tuner', *RadCom* September 2000.
[2] 'The Pic-A-Switch, a frequency dependent switch', *RadCom* September 2001.
[3] 'A low distortion sine-wave generator', *RadCom* February 2003.

# CHAPTER 6

# Remote control and telemetry

The areas of remote control and telemetry offer all sorts of possibilities to the radio amateur for using microcontrollers, and illustrate many of the hardware and software techniques applicable to other fields. Remote control is often used with repeaters and beacons operated on remote sites, and for this application various security measures have to be applied to ensure valid commands are issued and to prevent unauthorised users or 'hackers' from attacking the remote facility. Telemetry, or remote metering, is invaluable to permit parameters such as equipment status, supply voltage, cabin temperature, power output level or even door open/closed status to be monitored without having to make a site visit.

A range of PIC-based hardware designs is included here for these various functions but, more significantly, a library of PIC assembly code routines is developed, along with various algorithms where necessary, which can be transplanted and modified at will for any future project. This section also illustrates how PIC input and output functions can be expanded with additional hardware, such as shift registers to increase the number of I/O lines, higher-resolution A/D converters with more analogue ports, and external decoder chips.

## Dual-tone multifrequency signalling

The DTMF signalling scheme is now in worldwide use for telephone dialling. The digits from 0 to 9 plus two special characters, '*' and '#', are represented by two audio tones sent simultaneously – one tone is taken from a group of four corresponding to rows on a typical telephone keypad, the other one of three tones corresponds to the column. As an extension to the standard set, a fourth column and tone is allowed, permitting the letters 'A' to 'D' to be encoded and giving 16 unique codes. The tone frequency pairing with associated digit coding is shown in Table 6.1. Note that the code output from the chip for the '0' digit is actually the code for decimal '10' rather than zero. The zero code was originally adopted as an error condition and should never be generated from a valid tone pair but, when including the fourth column, this code now has to be allocated as a valid state in order to allow all 16 possible key combinations.

| Table 6.1. DTMF digit/tone coding | | | | | | |
|---|---|---|---|---|---|---|
| $F_{low}$ (Hz) | $F_{high}$ (Hz) | Digit | D3 | D2 | D1 | D0 |
| 697 | 1209 | 1 | 0 | 0 | 0 | 1 |
| 697 | 1336 | 2 | 0 | 0 | 1 | 0 |
| 697 | 1477 | 3 | 0 | 0 | 1 | 1 |
| 770 | 1209 | 4 | 0 | 1 | 0 | 0 |
| 770 | 1336 | 5 | 0 | 1 | 0 | 1 |
| 770 | 1477 | 6 | 0 | 1 | 1 | 0 |
| 852 | 1209 | 7 | 0 | 1 | 1 | 1 |
| 852 | 1336 | 8 | 1 | 0 | 0 | 0 |
| 852 | 1477 | 9 | 1 | 0 | 0 | 1 |
| 941 | 1209 | 0 | 1 | 0 | 1 | 0 |
| 941 | 1336 | * | 1 | 0 | 1 | 1 |
| 941 | 1477 | # | 1 | 1 | 0 | 0 |
| 697 | 1633 | A | 1 | 1 | 0 | 1 |
| 770 | 1633 | B | 1 | 1 | 1 | 0 |
| 852 | 1633 | C | 1 | 1 | 1 | 1 |
| 941 | 1633 | D | 0 | 0 | 0 | 0 |

With the widespread use made of DTMF coding, it is inevitable that a range of custom chips have become available to make use of this coding scheme. The decoding chips contain all the necessary filtering and decision circuitry, as well as detection and rejection of spurious/error tones. The final result is a very robust and error-free, albeit rather slow, digital coding scheme for voice-band channels which can be used on radio links just as easily as on telephone lines. One of these single-chip decoders that is straightforward to use is the Mitel MT8870D integrated DTMF receiver chip. With a handful of external resistors and capacitors, an audio signal input can be turned into a four-bit code presented on four parallel wires, with a fifth carrying a strobe signal. Mitel (now Zarlink), also produces other devices for DTMF coding, such as integrated transceiver chips. For the full range see reference [1].

## Telephone line remote controller

The unit described is intended for remote control of beacons and repeaters where a permanent telephone line is available for secure control. The unit may be used with a duplex radio link, for example when controlling a repeater, but for simplex (transmit/receive) operation, hardware and software changes will be required. Furthermore, issues such as security of control codes may inhibit use over a public radio channel.

On ringing the telephone number of the line to which the controller is connected, it will automatically answer after approximately two rings. A personal identity number (PIN) is then entered, followed by a command using DTMF tones generated by the calling telephone to control the remote circuits and read their status.

Direct connection to a telephone line requires that equipment be approved for such connection, so no complete details can be given here for such an interface. However, a suitable interface module for connection that may meet type-approval regulations is given at the end of the article. Some alternative means for making a safe connection to the Public Switched Telephone Network are also suggested. In all cases the relevant approvals and specifications should be read and followed in construction.

## Control protocol

Two levels of control are allowed. One set of PINs, intended for issuing to persons not holding an amateur radio licence, will turn OFF all controlled circuits with no ability to switch them back ON again using this number – the associated all-off commands are referred to as *Priority Commands*. Another PIN gives access to individual ON/OFF control of the remote circuits and allows the status of these, and the controller itself, to be read back using CW messages. This is referred to as the *User PIN*. If a Priority Command has been issued, this is indicated in the status message. Remote power supply monitoring is also featured, as the controller maintains a back-up battery to store the controlled circuits' state during power failures. If the beacon or repeater goes off-air this message allows the operator to see if power failure is the case or whether a command or equipment failure has occurred. The internal battery should be able to power the PIC and DTMF decoder for one to two days.

## Design

Fig 6.1 shows the circuit diagram of the controller but intentionally gives no details of interfacing to the telephone line – components within the box labelled 'Line Interface' are for illustration only. A dedicated MV8870 DTMF decoder IC performs all the audio filtering, validation and data decoding and sends a four-bit code representing the DTMF digit received, plus a strobe, to a PIC microcontroller. At all times the PIC is monitoring for valid DTMF codes, whether the interface is off-line or on-line. This allows for local on-hook signalling as well as non-phone line, or leased line use. The decoder chip requires a 3.58MHz clock generated by a crystal, and for convenience the same oscillator is also used to supply the PIC clock. The PINs are stored in the PIC's non-volatile memory at the time the PIC is programmed, and every DTMF sequence received is checked for a valid PIN. Up to three circuits can be controlled from this design and, as shown, two command outputs are in the form of a switch closure to ground intended for operating relays or similar, and one is an uncommitted 0/5V logic level.

The PIC generates audio CW messages used for acknowledging data entry and status messages. A LED is included to show locally when audio responses are being generated but is really there only as an aid during software development and is fully software programmable.

The final task of the PIC is to respond to the telephone line and perform the auto-answer function. An opto-isolator in the line interface monitors the line for ringing voltage and, when this is present, C8 is charged. When

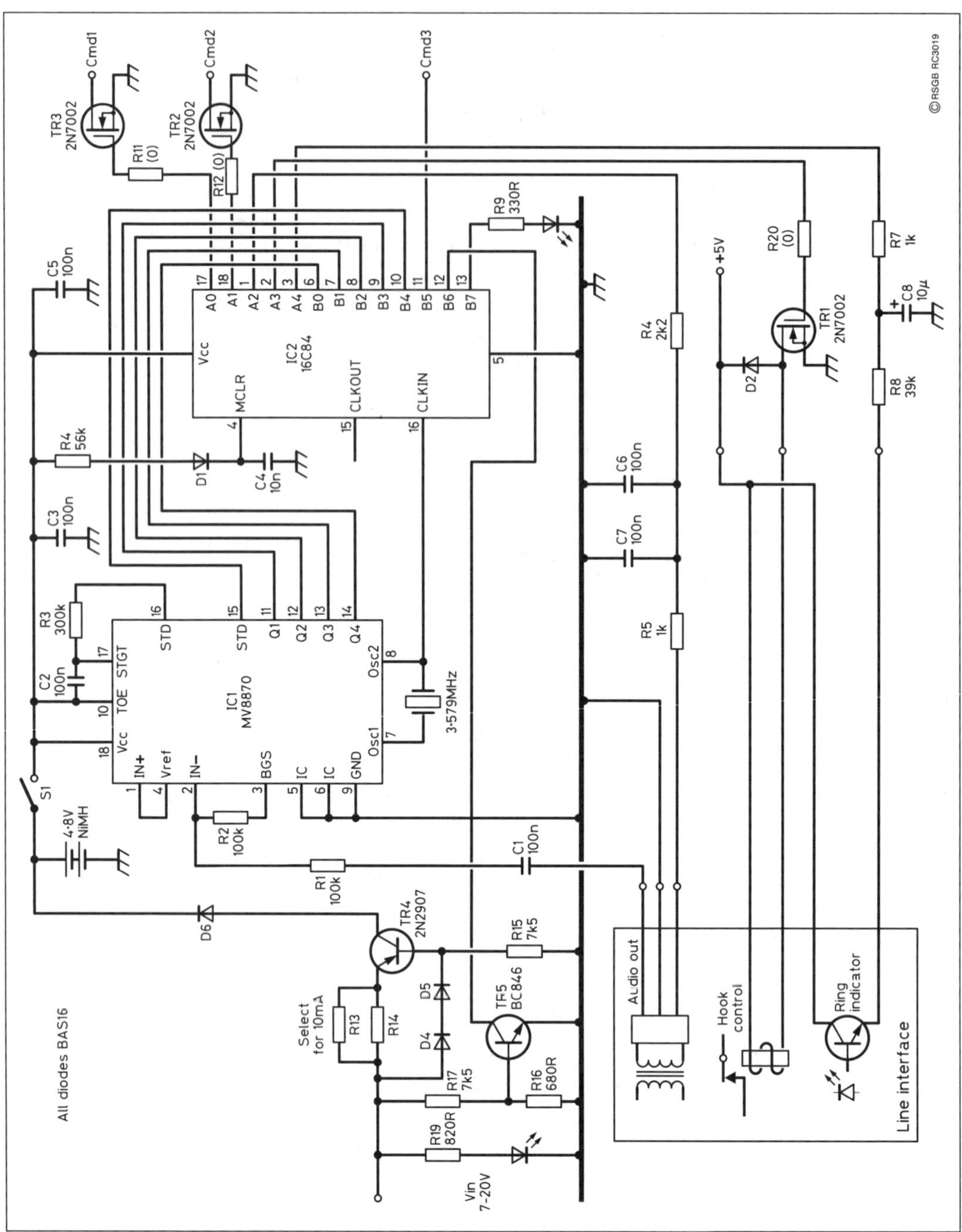


**Fig 6.1. Telephone line remote control circuit**

the switching threshold on the PIC's A4 Schmitt trigger input is reached, A3 is activated which operates the relay or switch in the line interface via Q1, seizing the line. C8 is then discharged, ready for the next call. To ensure fail-safe operation, the PIC software maintains a continuous time-driven interrupt counter. When a particular value of count is reached (after 20 seconds) the controller will release the phone line irrespective of the state of any command entry. Every time a DTMF digit is received the counter is reset to zero, restarting the 20s time delay. This means that is impossible to lock up the controller in a state that keeps the phone line latched on, and to terminate a control session just hanging up or entering no tones for 20s is all that is needed. Immediately before going off-line, a tone is sent.

The nominally 5V power supply for the MV8870 and the PIC is generated by a NiMH battery which is float charged by a constant-current source of 10mA. With a PIC current consumption of 6mA this leaves 4mA to keep the battery topped up. Voltage input can be anything from 7 to 20V.

## Telephone line interfacing

Telephone operators are very concerned that connections to their network cannot cause any harm to exchange signalling equipment, personnel or correct operation of the network. As part of European and world harmonisation, the requirement for individual approval of all such equipment is no longer needed (BABT approval) but design specifications and rules are in place which must be adhered to if direct connections are contemplated and approval by random selection of production items is considered satisfactory. Most of these are concerned with safety isolation, voltage breakdown and testing, but also such matters as audio drive levels and out-of-frequency band energy are specified. See reference [2] for more details.

Fortunately, ready-made modules are now available to perform all the line interfacing and safety barrier functions. Using of one of these means that when coupled with correct construction techniques and housing, a design can be made that is capable of meeting the regulations. No details will be given here as it is each builder's own responsibility to ensure that his or her construction meets the requirements. Many of the data sheets and application notes for telephone line interfacing components give helpful information on design and construction techniques – one of the most important is the need for at least 5mm creep distance over a PCB surface between any connection to the PSTN network and connections to the user's side.

## Using the remote controller

The remote control unit will auto-answer a telephone line and command three circuits on or off depending on DTMF codes sent to it. CW responses are generated showing the status of the controlled circuits.

Separate PINs are needed to access the controller. Three PINs are used individually to turn OFF each of the three controlled circuits. These are intended for issuing to non-licence holders to allow them to turn off beacon or repeater equipment without their having the ability to switch the equipment back on again. Receipt of any of these three PINs sets status flags that can be read by the user.

- PIN 1 turns OFF Circuit 1
- PIN 2 turns OFF Circuit 2
- PIN 3 turns OFF Circuit 2

- PIN 4 gives access to a command set which controls the individual ON/ OFF state of each command circuit and allows the controller's status to be read back in the form of CW messages. This PIN is intended for the beacon/repeater keeper(s) use only.

## Operation

After the controller's telephone number is dialled, it will answer after approximately two rings, and reply with a 1800Hz tone. At the end of this tone one of the PINs may be entered. Entering PIN 1–PIN 3 will immediately turn off its associated controlled circuit and respond with a double bleep to show the command has executed. If this is not heard, repeat by entering the full PIN again. PIN 1, PIN 2 etc may be entered in any order, waiting for the double bleep each time before entering the next PIN.

Alternatively, enter PIN 4 and after a short delay the controller will respond with a blip to say it is now waiting for a single command digit. If no blip is heard, re-enter the four-digit code. After hearing the response, enter a command from the list below and, depending on the command entered, the controller will respond with either a 'K' in Morse or the status messages followed by a 'K'. Acknowledgement is sent even if invalid command digits are entered – these entries will be ignored.

The command sequence can be repeated continuously, with PIN 4 + command entered each time. At any stage, if no response appears from a PIN or command entered, just repeat until accepted – problems can occur if tones are entered too quickly and before an acknowledgement for a previous entry is sent. On completion of sending commands, it is advisable to check the status with Command 7 before closing the link.

The single-digit commands are:

| | |
|---|---|
| 1 | Turns ON Circuit 1 |
| 2 | Turns ON Circuit 2 |
| 3 | Turns ON Circuit 3 |
| 4 | Turns OFF Circuit 1 |
| 5 | Turns OFF Circuit 2 |
| 6 | Turns OFF Circuit 3 |
| 7 | Replays the status of the controller – see below for status messages |
| 8 | Globally resets all the status flags |
| 9 | Read command counter |
| 0, #, * | Not used, ignored |

Status messages returned in CW are:

| | |
|---|---|
| 'X' | Priority Turn-off Command 1, PIN 1, has been issued. Cleared by User Command 1, 4 or 8 |
| 'Y' | Priority Turn-off Command 2, PIN 2, has been issued. Cleared by User Command 2, 5 or 8 |
| 'Z' | Priority Turn-off Command 3, PIN 3, has been issued. Cleared by User Command 3, 6 or 8 |

'1'     Circuit 1 is ON
'2'     Circuit 2 is ON
'3'     Circuit 3 is ON
'B'     Main power supply at the remote site is off (the controller includes its own local battery to maintain status and auto-answer)
'?'     Controller has been powered on and circuits *may* be in default/start-up states. Cleared only by issuing a Clear Status command (User Command 8)

X, Y and Z status message flags are cleared by setting (or clearing) their respective circuit, eg the X response indicating that Circuit 1 was turned off by a PIN 1 Priority 1 Command may be cleared by issuing a PIN 4 + 1, or 4, command to either turn on Circuit 1 or leave it switched off. They are also cleared by the Global Reset Status Command 8.

The command counter is a modulo 256 count of the number of active commands issued, ie after the count reaches 255 it rolls back to zero. Every time PINs 1–3 are sent, or User Commands 1, 2, 3, 4, 5, 6, 8 are issued, this counter is incremented by one; no increment is generated for commands that simply read back status or command count. The value is sent back as a CW message giving the hexadecimal value of the count. For example, a count of 15 will generate '0F' in CW; a count of 123 will generate the CW message '7B'. The counter cannot be reset other than by switching off the controller to reset it and allows multiple users to keep a track of commands sent.

There is a built-in delay between hearing a PIN or command and its response to allow use with one-piece telephones; units that mute the audio when sending tones may give problems if the un-mute delay is too long. The controller will send a 1800Hz tone and go off-line if no DTMF tones have been received for 20 seconds; there is no need to wait for this tone before hanging up. CW messages are sent with a 900Hz tone; multiple responses are sent in the order they appear in the list.

### Software

The complete source code listing is too long to list in its entirety, but Table 6.2 shows a few of the routines that are specific to this software, namely, retrieving data from the PIC user non-volatile EEPROM memory and the use of a stack in user memory to manipulate a series of bytes in sequence. None of the CW message coding is included in this software listing as it was implemented in a rather crude way, and a far more efficient CW coding scheme is given later. A complete listing, DTMFREM2, for the controller is available from the RSGB website.

# Remote control over an RF link

When using an RF link for remote control, several new problems now arise. The foremost of these is that as the RF link is not necessarily bi-directional, acknowledgement of a successfully received command is only supplied by

**Table 6.2. Some of the routines specific to the telephone remote controller PIC code**

```
CheckPin1                               ;looks for valid PIN1 in stack1-4
        movlw   0               ;Correct PIN kept in EEPROM locations 0-3
        call    GetEE           ;PIN sent MSB first, MSB saved in LOWEST
        subwf   Stack4 , W      ;EEPROM address so it looks correct in listing
        btfss   STATUS , Z      ;Separately check each digit in the stack against
        goto    NotValid1   ;  what is expected.
        movlw   1
        call    GetEE
        subwf   Stack3 , W
        btfss   STATUS , Z
        goto    NotValid1
        movlw   2
        call    GetEE
        subwf   Stack2 , W
        btfss   STATUS , Z
        goto    NotValid1
        movlw   3
        call    GetEE
        subwf   Stack1 , W
        btfss   STATUS , Z
        goto    NotValid1       ;If all four received digits check OK,
        movlw   1               ; return with 1 in W, else return 0 for error
        return
NotValid1
        movlw   0
        return
;————————————
PushStack                       ;Move new data into beginning of stack, and
        movf    Stack4 , W      ; push up all other data, throwing away the oldest.
        movwf   Stack5          ;Here the stack is 5 high, but could be extended
        movf    Stack3 , W
        movwf   Stack4
        movf    Stack2 , W
        movwf   Stack3
        movf    Stack1 , W
        movwf   Stack2
        return
;————————————
ClearStack                      ;Prevent spurious PIN entry from commands
        clrf    Stack1
        clrf    Stack2
        clrf    Stack3
        clrf    Stack4
        clrf    Stack5
;————————————
GetEE                       ;read byte from non-volatile EEPROM memory
        movwf   EEADR           ;enter with address to be read in W
        bsf     STATUS,RP0      ;ram page 1
        bsf     EECON1 , RD
        bcf     STATUS , RP0 ;ram page 0
        movf    EEDATA , W      ;Exit with data in W
        return
;————————————
        end
```

the controlled function, such as a beacon transmission, going off or on. If a transmitter and receiver pair is used at the remote site the link will still probably be simplex and commands cannot be verified as soon as they are issued.

A solution to this is to add an input to the controller monitoring the receiver squelch, so that command acknowledgement is sent only when the transmitting station drops carrier. Another output from the controller is needed to operate the transmit/receive line. This assumes an FM or AM link is used, which is usually necessary for DTMF coding to ensure proper frequency accuracy. One issue, still to be resolved with this system, is coping with what happens if the squelch is being held open by a weak interfering signal that does not affect the command transmission. The solution adopted is to wait for a suitable time – say 10 to 20 seconds – for the squelch line to drop. If it doesn't close then the acknowledgement transmission is sent anyway, and hopefully will be received by the controlling station who will now be alerted (by the delay) that the channel may have some interference present.

## Link security and coding issues

The next issue is one of security. Any casual listener can listen in to the commands being sent, and if the same PIN is used each time it is easy to monitor these and decode the digits, so allowing unauthorised access and the possibility of false or malicious commands being issued. The solution is to adopt a scheme where the PIN changes for each access – only authorised users have the list of which valid PINs will be accepted each time. A similar scheme is used for modern automobile door locks and immobilisers, where a small radio-operated keypad transmits a continuously changed message that is known only to the controller each time a button is pressed to permit entry.

A pseudo-random sequence can be used to generate a series of PINs, provided the method of generating the sequence is sufficiently concealed so that a casual interceptor cannot predict which ones will occur next time. We met PRN sequences in the last chapter where a random sequence of data was generated from a feedback shift register combination. The requirement here is broadly similar, except that a range of numbers is now needed instead of a single bit changing. The sequence needs to be much longer to prevent the algorithm being worked out by an interceptor, and it has to be possible to easily generate different sequences for different on-air links or users.

A convenient way of generating pseudo-random numbers with simple mathematics is by use of the equation:

$$N = (A.N + B) \bmod C$$

Each new value of $N$ is generated by multiplying its previous value by $A$, adding $B$ and taking the modulus to $C$ of the result (the remainder left after repeatedly subtracting $C$) so the resulting value of $N$ can range from zero to a maximum of $C - 1$. $A$, $B$ and $C$ have to be ideally chosen so that the sequence is long enough and has no repeats. However, even for arithmetic limited to 16-bit integers a wide range of possible values are available, and 16-bit maths is reasonably straightforward to write in PIC assembler code. The maximum length of the generated sequence cannot be any longer than $C$ before it repeats, and for certain choices of $A$, $B$ and $C$ can be a lot shorter. To ensure maximum length it is advisable to check the sequence generation by modelling it in software first – in any case this step will be needed to generate the list of PINs for each user and a spreadsheet programme is adequate. As the modulus function is not straightforward to programme in assembler code from first principles, the value for $C$ can often conveniently be made a power of two, such as $2^{13} = 8192$, so taking the modulus now becomes a case of just extracting the correct number of lowest significant bits of the result from the multiply-and-add calculation.

As an example, using the values $A = 17$, $B = 799$ and $C = 256$, the first few values starting with an initial 'seed' value for $N$ of 1 are 1, 48, 79, 94, 93, 76, 43, 250, 185 etc which certainly look at first sight as if they may be random.

Altering any of $A$, $B$ or $C$ results in a widely differing sequence, especially after the first few terms, and the values $A$, $B$ and $C$ are sometimes called the *key* when used in encryption. This technique of random number generation is used within many programming languages – in Basic it is invoked by the RND function.

To generate such a sequence on a PIC requires a multiplication routine. The latest high-range PICs, the 18Fxxx family, include an $8 \times 8$-bit hardware multiplier which gives a 16-bit result and can be cascaded for larger numbers. However, a multiplication routine has to be made up for the smaller microcontrollers. In binary, traditional long multiplication is reduced to a series of multiply by two with conditional add instructions. In the sum $A \times B$ for example, the value of $B$ is repeatedly multiplied by 2, which is the same as shifting it left by one bit each time, and if the corresponding bit in $A$ is set, accumulating the result as illustrated in the 'Binary multiplication' box opposite.

The listing in Table 6.3 is for a subroutine that generates the next pseudo-random number in a sequence from its previous value, within the range 0 to 8191. The constants to be used for the calculation are limited to an eight-bit value for $A$ (stored in the register Avalue) and 16 bits for $B$ in the two registers Bhi/Blo. The previous value, to be modified, is passed into the subroutine with its most significant bits in the register Whi and the lowest bits in the working W register. The value of $C$ is made equal to 8192 and the modulus, or truncation, of the sum is done by taking only the lowest five bits of the higher-order byte of the result. Note that with the values used in this sequence, the register AccHi is not needed, although it is included within the software listing for completeness.

## Expanding I/O ports

Often more output functions need to be controlled than there are available pins on the controller. One solution is of course to use one of the functionally

<table>
<tr><td colspan="3" align="center">Binary multiplication</td></tr>
</table>

A = 147 or in binary 10010011
B = 42 or in binary 00101010

We also need a working register, usually called an *accumulator*, X. We will shift B, while testing the bits in A, starting with the lowest significant ones, and depending on the result of the test decide whether to add the shifted B value into X or not.

Bit 0 of A = 1, so set the working accumulator, X = B.

Now shift B left, so multiplying by 2. Now B = 1010100.

Bit 1 of A is also 1, so this new B value is added into the accumulator, X = X + 2.B or X = B + 2.B

Shift B left again, multiplying the original value by 4, B = 10101000.

Now bit 2 of A is a zero so we do nothing and continue. After the next shift bit 3 of A is also zero so we continue again.

After the fourth shift B = 1010100000 and is now 16 times its original value. At this stage, Bit 4 of A is again set, so this new value of B is added into the accumulator. Now X = B + 2.B + 16.B

The next two shifts are not added to the accumulator as bits 5 and 6 of A are zero, but the final shifted result of B, equal to 128 × the original value, is included so now X = (1 + 2 + 16 + 128) × B.

This process is illustrated below.

```
A              X
1        10 1010     B
1       101 0100     B × 2
0
0
1     10 1010 0000   B × 16
0
0
1   1 0101 0000 0000 B × 128
    ________________________
    1 1000 0001 1110 B × (128 + 16 + 2 + 1)
```

From this it is clear that not only does a 16-bit register (the accumulator) have to be allocated to contain the results of an 8 × 8-bit multiplication but the B value also has to be stored as 16 bits to allow it to be shifted at each stage. Furthermore, the value of B in its original form is destroyed, or at least modified, so if it is needed later this has to be saved separately.

similar devices that has more I/O registers and pins. However, these generally cost rather more than the smaller devices, and still require interfacing from the 0/5V level to whatever is needed to drive the controlled equipment.

A device that simplifies interfacing to relays and other higher-powered loads is a serially loaded driver chip such as the UC5841. These typically have a serial input where data is clocked into a shift register sequentially via data and clock lines, then stored in output latches by a load pulse on a third

**Table 6.3. Pseudo-random sequence generation**

```
NextCode          ;Takes in Whi/W, uses Avalue & Bhi/Bl, calculates new Whi/W
        movwf   Temp              ;Whi/Temp  is register to shift & add

        clrf    AccLo
        clrf    AccMe
        clrf    AccHi
        movlw   8                 ;Multiplicand limited to 8 bit value
        movwf   Counter
MultLoop
        rrf     Avalue            ;Working bit into carry
        btfss   STATUS , C
        goto    NoMult

        movf    Temp , W          ;get shifted Wlo value
        addwf   AccLo             ; and add into accumulator if 1 in
        btfss   STATUS , C        ; appropriate bit in Avalue
        goto    Add1
        incf    AccMe
        btfsc   STATUS , Z
        incf    AccHi
Add1
        movf    Whi , W
        addwf   AccMe
        btfsc   STATUS , C
        incf    AccHi
NoMult
        bcf     STATUS , C
        rlf     Temp          ;Multiply by 2 ready for next partial product
        rlf     Whi
        decfsz  Counter
        goto    MultLoop
                              ;ACC = A.W
        movf    Blo , W
        addwf   AccLo
        btfsc   STATUS , C
        incf    AccMe
        btfsc   STATUS , C
        incf    AccHi
        movf    Bhi , W
        addwf   AccMe
        btfsc   STATUS , C
        incf    AccHi             ;ACC = A.W + B

        movlw   0x1F
        andwf   AccMe             ;AccMe/Lo = (A.W + B) MOD 8192

        movf    AccMe , W
        movwf   Whi
        movf    AccLo , W         ;W/Whi = (A.W + B) MOD 8192
        return
```

line. The chips incorporate high current drivers of up to several hundreds of milliamps capability, which can operate at high voltages for directly driving

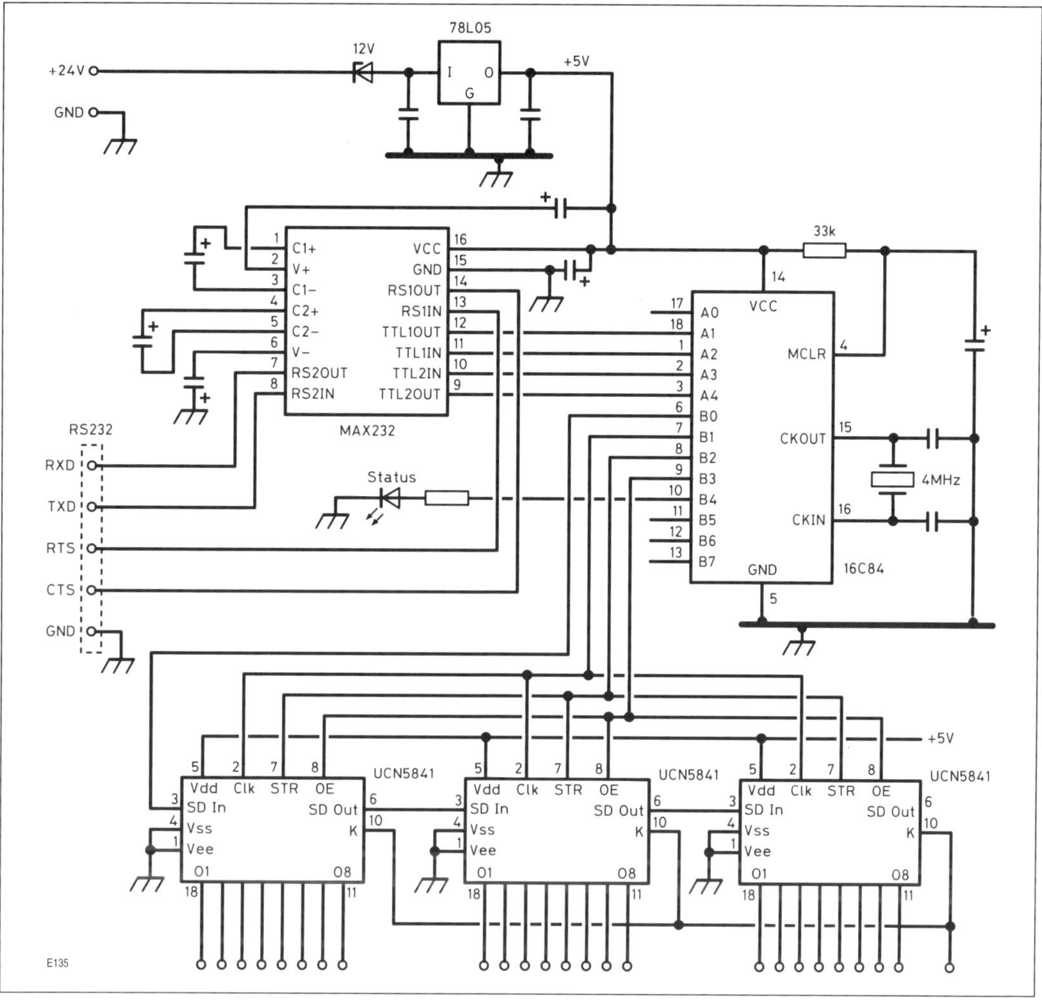


Fig 6.2. Using serially loaded relay driver chips

relays. They usually contain back-EMF diodes for protection from the transients generated by switching the inductive relay coils. Compared with the relay driver circuit shown in Fig 4.3, five more I/O lines are now released for other interfacing functions and, if more outputs are needed, the driver chips include a serial output from the final stage for cascading multiple devices. In theory there is no limit to the number that can be cascaded, but as the entire data set has to be serially clocked in each time a single relay is switched, 32 or 40 individual outputs is probably a sensible upper limit.

Using three of these devices, Fig 6.2 shows the circuit diagram of an interface for controlling up to 24 relays or other DC switched loads. Table 6.4 is a subroutine for the 16Fxxx family of PIC devices for transferring the contents of three data storage registers across the serial interface to a trio of cascaded UC5841 chips.

Fig 6.3 shows the circuit diagram for a complete telecommand system for switching up to eight circuits at a remote facility, making use of the concepts

**Table 6.4. Code for sending data to serial driver chips**

```
;Data stored in D1/D2/D3
;Output lines to UC5841 devices, Dout, Clk and Strobe

SendRelayData                          ;MSB first for UCN5841 relay driver
        movlw   d'24'                  ;24 bits, data clocked in MSB first
        movwf   Counter                ;Data stored in D3/D2/D1
RelayLoop
        bcf     Clk                    ;data changes with -ve edge of clock
        rlf     D1                     ;Data transferred D3MSB >> D1LSB
        rlf     D2
        rlf     D3              ;
        btfss   STATUS , C              ;Test and transfer to Dout line
        bcf     Dout
        btfsc   STATUS , C
        bsf     Dout
        bsf     Clk                     ;stobe in data bit
        nop                           ;Short delay to lengthen clock pulse
        nop
        decfsz  Counter                 ;do this 24 times
        goto    RelayLoop
        rlf     D1      ;Restore original pattern and preserve registers
        rlf     D2
        rlf     D3
        bcf     Clk
        nop
        nop
        bsf     Strobe      ;now strobe in the entire data set
        nop
        nop
        bcf     Strobe
        return
```

outlined so far. DTMF tones as generated by the majority of 144 and 432MHz handi-talkies are used to allow the operator to enter the digits. Three completely separate PIN sequences are maintained so that up to three users can independently gain access to the control system without having to inform each other of the position in the sequence. A readback function is included to allow users to read the status of the eight controlled circuits as well as the current position in each of the three code sequences. This latter function enables each user to keep track of the number of commands that may have been issued by the other authorised operators.

The software listing is too long to show here, but a complete listing is available from the website under the name 'DTMFRAD'. Note that the *A*, *B* and *C* values for the three PIN sequences included in this listing are *not* those actually used for any operational version of this system!

## Telemetry

Remote monitoring or telemetry usually involves measurement of analogue parameters such as voltage, current consumption and temperature, and also

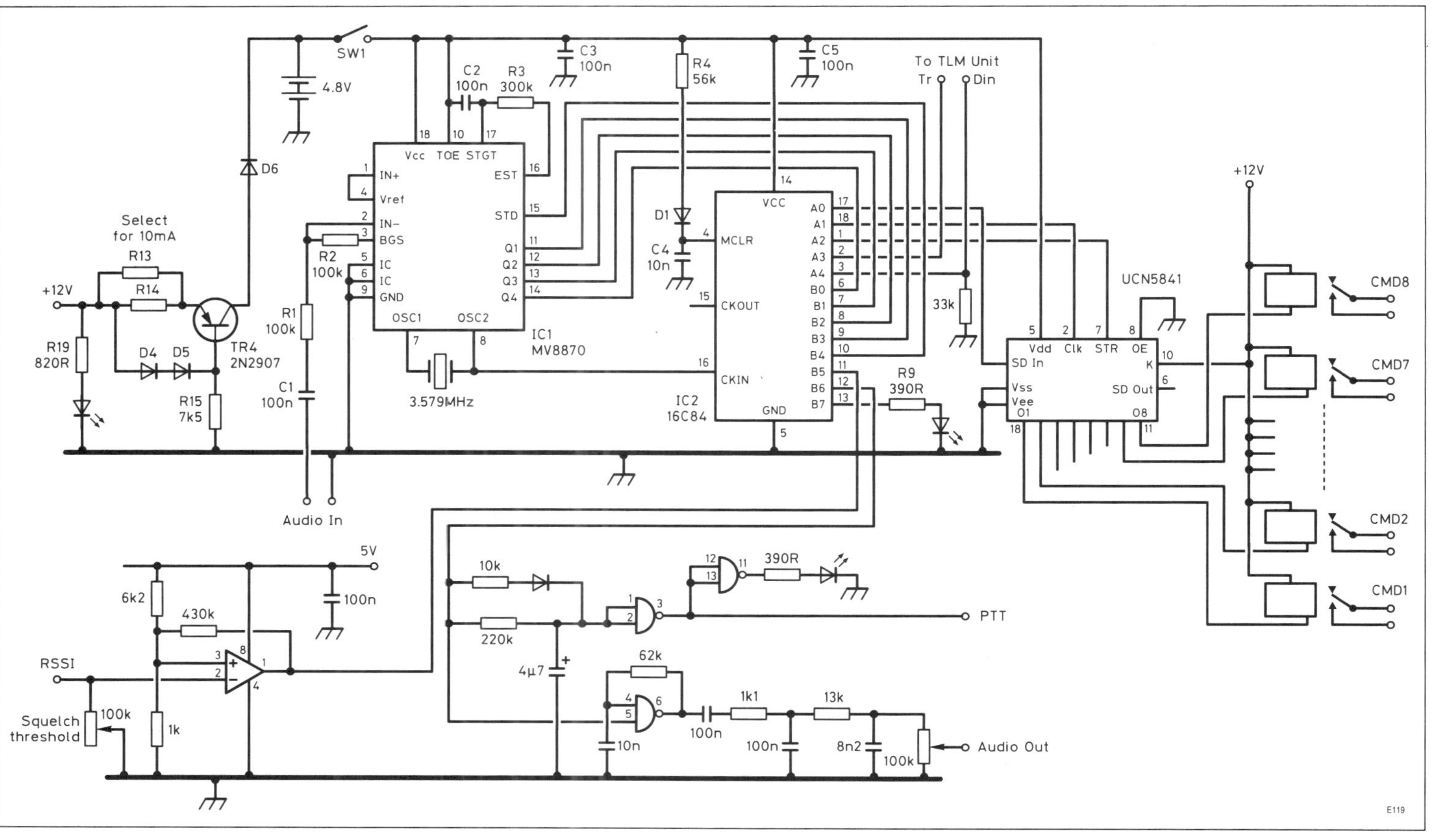

**Fig 6.3. DTMF controlled telecommand interface with transceiver interface**

the status of various items such as door and window switches as part of site security, equipment functional status etc. How the data is returned to the operator is also significant. While DTMF tones make commands easy to generate, using these on the return link is not so convenient. Data transmission is probably the most versatile system, but will normally require a PC to decode and display the received data. For a few parameters, returning these as automatically generated CW messages has a certain attraction, as a handi-talkie can continue to be used for both command generation and telemetry monitoring.

But first, some additional functions and hardware have to be added around a PIC to turn it into a useful telemetry system. Fig 6.4 shows the circuit diagram of a remote telemetry interface that operates in conjunction with the command interface of Fig 6.3. This telemetry module was designed to monitor AC voltage and current of the mains input to the GB3SCx microwave beacons installed on a remote hill in Dorset. Parameters that required to be monitored remotely were the mains voltage and current supply to the cabin, DC battery float-charge voltage, shack temperature and the status of two door switches. Some spare capacity is included in the telemetry hardware for further expansion. The AC power interface produces fully isolated 0 to 5V signals for the A/D converters, with full scale corresponding to an RMS current of 2.5A and voltage of 250V.

Temperature is measured via a thermistor and resistor, the values of the resistor and positioning on the 0–5V scale being determined experimentally to give an approximately linear voltage swing with temperature, giving an accuracy better than $2°$ in the range 0–30°. A better solution, giving a more accurate reading, would be to use one of the temperature measurement ICs such as the LM35 family of devices that give a linear voltage proportional to temperature, but the thermistor was already to hand and only required a two-wire interface, so it was used here.

### External A/D converters

Serial A/D converter chips have already been mentioned in Chapter 4, and here the use of a typical device is described; the four-wire interface to the PIC processor can be seen in Fig 6.4. The MC3208 device is a 12-bit converter with eight input channels which can be selected individually by an internal analogue multiplexer, so that up to eight separate voltages can be measured to a resolution of 5/4096 or approximately 1.25mV. A full data sheet for this and similar devices can be obtained from reference [3]. Op-amp unity gain buffers are included on the two external inputs from the AC mains interface to provide an element of EMC and protection against transients, working in conjunction with the RC networks on the input lines. All voltages are scaled to be within the range 0 to 5V, suitable for the A/D converter.

Table 6.5 lists the assembler code subroutine needed to read one channel from this device, the channel to be read being specified by loading it into the W register before calling the routine. First the set-up information, including channel number, has to be clocked into the A/D chip, and this device takes up four interface lines to allow this data input. After the conversion is

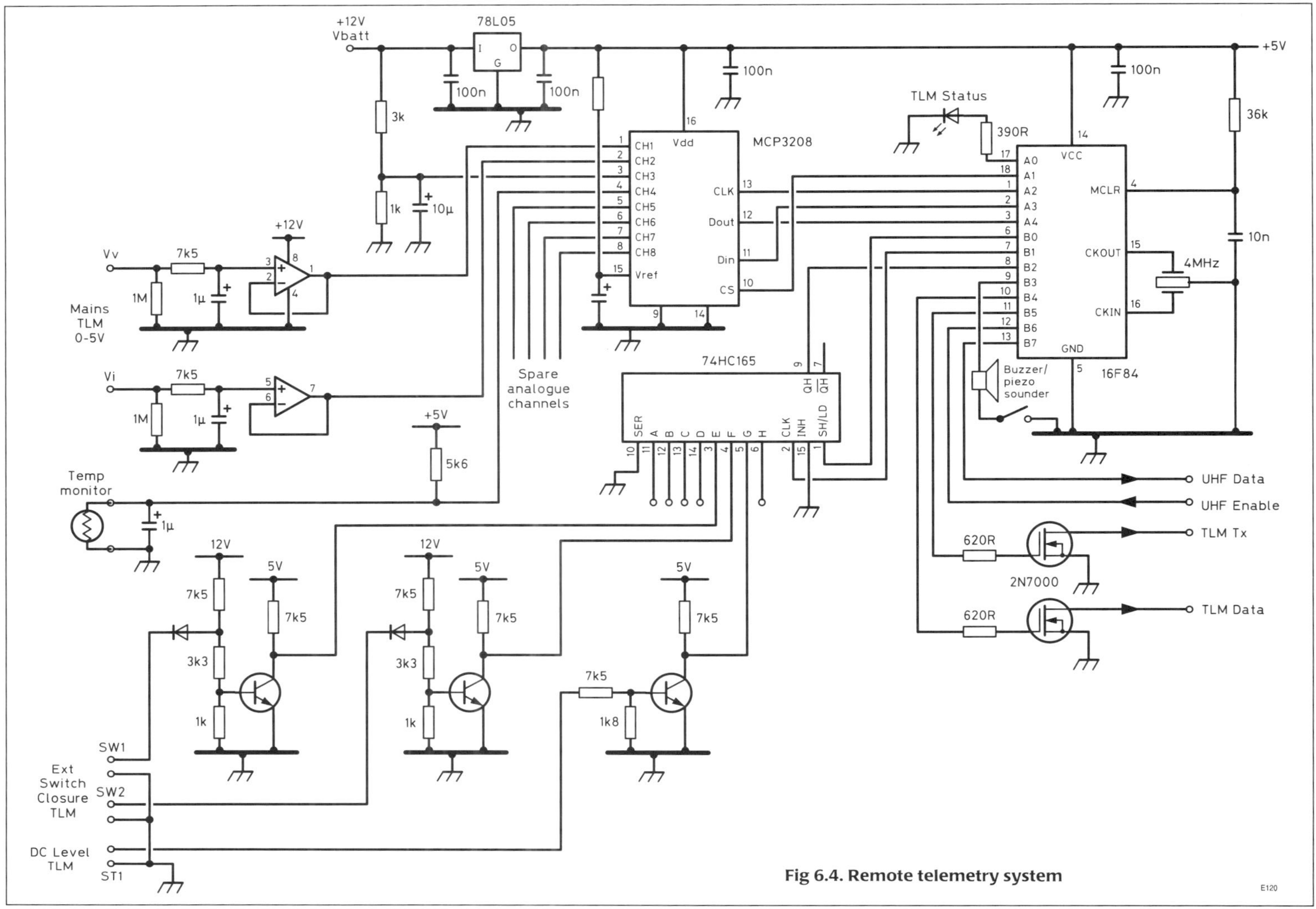


**Fig 6.4. Remote telemetry system**

**Table 6.5. PIC routine to read from external MC3208 12-bit serial, eight-channel A/D converter chip**

```
;Reads MCP3208 A/D converter
;Uses registers Alo , Ahi , Counter, Temp and lines ADCS, ADCLK, ADIN, ADOUT
ReadAD
        movwf   Temp                ;W contains channel to read
        rlf     Temp
        rlf     Temp
        rlf     Temp
        rlf     Temp
        rlf     Temp        ;Position channel no. at MS end, ready to shift out
        clrf    Alo         ;  MSB first, else 4 bits of last reading remain in MSBs
        clrf    Ahi
        bcf     ADCLK           ;Just in case
        bcf     ADCS            ;Enables device
        bsf     ADIN            ;Start bit
        bsf     ADCLK
        bcf     ADCLK
                            ;First clock in setup information
        bsf     ADIN            ;Single ended mode
        bsf     ADCLK
        bcf     ADCLK
        movlw   3               ;Now clock in channel no.
        movwf   Counter
ADChanLoop
        rlf     Temp
        btfsc   STATUS , C
        bsf     ADIN            ;Terminology 'IN' & 'OUT' refer to
        btfss   STATUS , C      ;   MCP3208 pin names !
        bcf     ADIN
        bsf     ADCLK
        bcf     ADCLK
        decfsz  Counter
        goto    ADChanLoop
        bsf     ADCLK           ;Trigger Sample & Hold, ADIN don't care
        nop
        bcf     ADCLK           ;This edge triggers output of B11
        nop
        bsf     ADCLK           ;Dummy read null bit
        nop
        bcf     ADCLK           ;Data MSBit wil appear now
        movlw   d'12'            :Now serially clock out the 12 data bits
        movwf   Counter
ADLoop
        bsf     ADCLK           ;MSB 11
        bcf     STATUS , C      ;Read data bit into carry
        btfsc   ADOUT
        bsf     STATUS , C
        bcf     ADCLK           ;Triggers O/P of next bit
        rlf     Alo             ;Rotate into AD Data registers
        rlf     Ahi
        decfsz  Counter
        goto    ADLoop
        bsf     ADCS            ;Returns A/D  to standby
        return
```

**Table 6.6    Routine to read 8 status bits via 74HC165 shift register interface**

```
;Shift register connected: Load - StsLoad,  Clock - SerClk,  Serial Data out - StsData

ReadStBits                    ;Get data from  shift register, returned in LineData
    bcf      StsLoad            ;Strobe in parallel data on -ve edge
    nop
    nop
    bsf      StsLoad          ;First bit appears at StsData immediately after Load pulse
    clrf     LineData
    movlw    d'8'
    movwf    Counter
ShRegLoop8
    btfsc    StsData
    bsf      STATUS , C
    btfss    StsData
    bcf      STATUS , C
    rlf      LineData

    bsf      SerClk           ;Data changes on +ve clock edge
    nop                         ;Clock through next bit of data
    nop
    bcf      SerClk

    decfsz   Counter      ;Until all 8 bits are done
    goto     ShRegLoop8
    return
```

triggered, the data is clocked out one bit at a time, starting with the MSB into
the PIC registers Ahi and Alo. Interestingly, and unlike the A/D converter
within some members of the PIC family, there is no need to wait for the
conversion to complete before reading out the result. Most A/D converters
work on a successive approximation principle, where the reading is generated one bit at a time starting with the MSB. Hence, when clocking data out
serially, each bit can be read as soon as it is generated. Some A/D chips allow
the option of clocking out data LSB first, and in this instance it is necessary
first to wait for conversion to complete.

## Status telemetry

Referring again to Fig 6.4, a 74HC165 shift register is included to enable up
to eight on/off status lines to be read, while only tying up three of the PIC
I/O pins. Two of the shift register inputs are connected through transistor
buffers to switches mounted on doors and windows. The buffers are there
to give some isolation against short-circuits and transients that might otherwise get passed through to the logic circuitry. Another input comes via a
transistor buffer and is designed to monitor a status voltage from a separate
area of the beacon system. If more than eight bits are wanted, a second or
further 74HC165 device(s) can be cascaded by connecting the final stage
output to the serial input of the next one, then connecting the load and
clock pins together. The software then has to be changed to clock out the
necessary number of bits into a set of registers to hold the data.

### Transmitting the data

This telemetry system was designed to operate as a stand-alone unit, sending data that could be easily read by any operator with no more than a UHF handheld radio. The outputs of the A/D converter, consisting of a binary number from 0 to 4095 and representing 0–5V input for each channel, have to be scaled to give the correct value as a decimal number for the appropriate parameter being measured, and converted into decimal for transmission.

Each of the four analogue parameters has a different scaling factor depending on the potential divider or other analogue network used to generate the 0–5V input signal and in most cases a straightforward scaling from the 0–4095 value is all that is needed, ie multiplication or division of this number by a constant. As an example we will take channel number 02 which measures the DC battery float charge voltage. Table 6.7 lists the portion of code that performs this function. With 12 bits of measurement resolution available, more than sufficient accuracy can be obtained by simply scaling the voltage by a factor of 0.25, so the 0–5V input range corresponds to an actual allowed DC voltage range of 0–20V, more than enough to show the battery condition. So now a 0–20V signal generates a code from 0–4095 which has to be converted to binary coded decimal (BCD) numbers ready for transmission. The scaling factor is therefore equivalent to multiplying the value $N$ from the A/D converter by 20/4096, or dividing by 204.8. We saw earlier that multiplication on a PIC requires a special routine to perform, and then only integer arithmetic is possible. By employing some crafty tricks we can simulate decimal fractions. The technique used is to scale to a value of 0 to 2000 then just put in a decimal point at the correct place; the scaling factor now becomes $2000/4096 = 250/512$.

We already have a routine for multiplication, although that listed in Table 6.3 needs to be changed slightly to accept the 12-bit values used here. Division, however, is a more complex routine to write from first principles, involving repeated shift and subtract, and will usually give non-integer results; here we restrict division to a binary shift only meaning, for example, that a number can be divided by 512 by shifting the result nine bits to the right. A generalised routine for binary division by $2^N$ is shown in the Div2N routine listed, so now the scaling operation for the DC supply voltage is reduced to a single multiplication by 250 followed by a nine-bit right shift.

The other parameters are scaled appropriately:

$$\text{Channel 0} \quad 0\text{–}250\text{Vac} \times 625/1024 = 0\text{–}2500$$
$$\text{Channel 1} \quad 0\text{–}2.5\text{Aac} \times 625/1024 = 0\text{–}2500$$

Channel 3, the temperature, is rather more complicated but the following equation was determined experimentally:

$$T = (3000 - A) \times 51/2048$$

This gives a temperature reading accurate to a couple of degrees within the range of at least 0–30°.

We now have a binary number stored in two bytes which then has to be

**Table 6.7. PIC code to read a voltage and convert to decimal values**

```
        "
        "
        "
;Channel 2 Data      DC Supply 0 - 20V
        movlw    2
        call     ReadAD    ;A/D channel 2

        movlw    0x00
        movwf    Bhi
        movlw    d'250'    ;Multiplicand = 250
        movwf    Blo

        call     Multiply    ;Acc = A * 250
        movlw    d'9'
        call     Div2N         ;Acc = A * 250 / 512 = 0 - 2000 = 20.00 v
        movf     AccMe , W
        movwf    Ahi
        movf     AccLo , W
        movwf    Alo
        call     BinToBCD

        movf     BCDMe , W
        call     CWHiNibble
        movf     BCDMe , W
        call     CWLoNibble
        movlw    "E"
        call     SendCW
        movf     BCDLo , W
        call     CWHiNibble
        movf     BCDLo , W
        call     CWLoNibble
        movlw    "V"
        call     SendCW
        "
        "
        "
;Main code continues…….


;───────────────────
Multiply                ;Ahi/Alo * Bhi/Blo, result in AccLo/AccMe/Acchi
        clrf     AccLo    ; A & B destroyed   , max 24 bit product
        clrf     AccMe
        clrf     AccHi         ;Clear accumulators
        clrf     Temp          ;Temp will contain MSBs of shifted B
        movlw    d'12'         ;A and B limited to 12 bit values only.
        movwf    Counter
MultLoop
        bcf      STATUS , C
        rrf      Ahi           ;Working bit into carry
        rrf      Alo           ;Repeatedly adds shifted version of B
        btfss    STATUS , C    ;  into Acc depending on working bit of A
        goto     NoMult

        movf     Blo , W       ;get shifted Blo value
```

**Table 6.7 *(continued)***

```
          addwf   AccLo                ; and add into accumulator if 1 in
          btfss   STATUS , C           ; appropriate bit in Avalue
          goto    Add1
          incf    AccMe
          btfss   STATUS , Z
          goto    Add1
          incf    AccHi
Add1
          movf    Bhi , W
          addwf   AccMe
          btfss   STATUS , C
          goto    Add2
          incf    AccHi
Add2
          movf    Temp , W
          addwf   AccHi
NoMult
          bcf     STATUS , C
          rlf     Blo                  ;B = B * 2 ready for next partial product
          rlf     Bhi
          rlf     Temp
          decfsz  Counter
          goto    MultLoop

                                       ;AccHi/Me/Lo = A * B
          return
;———————————————
Div2N                                  ;Divides Acc by 2^N.  N in W
          movwf   Counter              ;  nb. Does not work if entered with 0
DivLoop
          bcf     STATUS , C
          rrf     AccHi
          rrf     AccMe
          rrf     AccLo
          decfsz  Counter
          goto    DivLoop
          return


BinToBCD            ;Takes in Ahi/Alo, calculates packed BCD equivalent
          clrf    BCDHi            ;also uses Temp, Counter, destroys Ahi/lo
          clrf    BCDMe
          clrf    BCDLo
          movlw   d'16'
          movwf   Counter
BcdLoop
          rlf     Alo
          rlf     Ahi
          rlf     BCDLo
          rlf     BCDMe
          rlf     BCDHi
          decfsz  Counter          ;Need a final shift only, so loop count here
          goto    BCDgo
          goto    BCDone
   ;..........
BCDgo
```

**Table 6.7** *(continued)*

```
        movf    BCDHi , W
        addlw   3
        movwf   Temp
        btfsc   Temp , 3
        movwf   BCDHi        ;If BCDHI >= 5, add 3 and store back

        movf    BCDHi , W
        addlw   0x30
        movwf   Temp
        btfsc   Temp , 7
        movwf   BCDHi        ;Adjust BCDHI if > 80

        movf    BCDMe , W    ;Repeating for middle then low digits
        addlw   3
        movwf   Temp
        btfsc   Temp , 3
        movwf   BCDMe

        movf    BCDMe , W
        addlw   0x30
        movwf   Temp
        btfsc   Temp , 7
        movwf   BCDMe

        movf    BCDLo , W    ;
        addlw   3
        movwf   Temp
        btfsc   Temp , 3
        movwf   BCDLo

        movf    BCDLo , W
        addlw   0x30
        movwf   Temp
        btfsc   Temp , 7
        movwf   BCDLo

        goto    BcdLoop
;.................
BCDone
        return
```

converted to a form readable by an operator. This is the function of the routine BinToBCD shown in Table 6.7 which takes in a 16-bit binary value and generates five BCD digits packed two at a time into the registers BCDHi/Me/Lo.

This is not an easy routine to describe, but is one that has appeared in many computing journals and articles over the years as an example of the 'best' way to undertake this rather difficult conversion. The easiest way to work out how it functions is simply to start with a large piece of paper and a few trial numbers, then work through the calculation one bit at a time – it really does work!

## Table 6.8. Compact CW character storage and message coding

```
SendCW                          ;Sends a single CW character stored in W
                                ;Keys BUZZ line to activate audio tone
        bcf     GieFlag
        btfsc   INTCON , GIE    ;Inhibit all interrupts to avoid corrupting
        bsf     GieFlag         ;  CW data out, but save its status for return
        bcf     INTCON , GIE    ;

        addlw   0xE0            ;Enter with ASCII character in W
        call    CWTable         ;Subtract 32 (add 256 - 32) to point to table entry
        movwf   CWChar          ;RETLW Doesn't affect STATUS, so check for zero separately
        movf    CWChar          ;  check for invalid returned code here
        btfsc   STATUS , Z
        goto    NoChar

        movlw   b'00000111'
        andwf   CWChar , W      ;Extract element count into W
        movwf   Counter

        movlw   6
        subwf   Counter , W     ;W = Elcount - 6. If +ve/0 then C=1, so > 5 els
        btfss   STATUS , C
        goto    Send5

        rrf     Counter         ;Puts LSB (first symbol) into C as 110 = .  111 = -
        call    SendEle         ;First  . or -  for 6 element characters
        movlw   5
        movwf   Counter         ;Force count to 5 for rest of the elements
Send5
        rlf     CWChar          ;Element to be sent into C.
        call    SendEle         ;   count already in Counter
        decfsz  Counter
        goto    Send5
NoChar                          ;Inter word gap, and invalid codes.
        call    CWDelay
        call    CWDelay         ;

        btfsc   GieFlag
        bsf     INTCON , GIE    ;Re-enable Timer interrupt if appropriate

        return
;-----------
SendEle                         ;Data in STATUS , C;  0 = dit  1 = dah
        bsf     BUZZ            ;Enable audio
        btfss   STATUS , C      ;test for dot or dash
        goto    SendDot
        call    CWDelay
        call    CWDelay
SendDot
        call    CWDelay
        bcf     BUZZ            ;End tone
        call    CWDelay         ;Inter element gap
        return
;-----------
CWDelay
```

**Table 6.8 *(continued)***

```
        movlw   d'84'       ;Adjust this value to get suitable dot period
        movwf   DelCount2
CWDelLoop
        call    Delay1ms
        decfsz  DelCount2
        goto    CWDelLoop
        return
;---------------
Delay1ms                        ;Cycle time = 1us
        movlw   d'197'          ;Delay = 5.N + 12 =  997us
        movwf   DelCount        ;  + 3us in higher loops = N . 1ms
BasicDLoop
        nop
        nop
        decfsz  DelCount
        goto    BasicDLoop
        return
;---------------------
org     0x3A0  ;Force tables to end to avoid page boundary problems.
CWTable               ;  CW Characters in compressed form.
        movwf   Temp      ;Set page register
        movlw   3
        movwf   PCLATH        ;
        movf    Temp , W    ;Recover offset
        addwf   PCL         ;Generate table address.
        ;       Binary pattern  Char  Hex
        retlw   b'00000000'  ; [sp]  00  Space also used for invalid characters
        retlw   b'00000000'  ;  !    00
        retlw   b'10010110'  ;  "    96
        retlw   b'00010101'  ;  #    15
        retlw   b'00000000'  ;  $    00
        retlw   b'00000000'  ;  %    00
        retlw   b'00000000'  ;  &    00
        retlw   b'11110110'  ;  '    F6
        retlw   b'10110101'  ;  (    B5
        retlw   b'01101111'  ;  )    6F
        retlw   b'01000101'  ;  *    45
        retlw   b'00000000'  ;  +    00
        retlw   b'10011111'  ;  ,    9F    Coding for CW characters of up to
        retlw   b'00001111'  ;  -    0F      6 elements.  Stored as :
        retlw   b'10101110'  ;  .    AE    Bits 7 to 3 make first five CW elements
        retlw   b'10010101'  ;  /    95    Reading  Left to Right
        retlw   b'11111101'  ;  0    FD    Coded as 0 = dit   1 = dah
        retlw   b'01111101'  ;  1    7D    Bits 2-0 are number of elements except for
        retlw   b'00111101'  ;  2    3D    110  1st element is a dit
        retlw   b'00011101'  ;  3    1D    111  1st element is a dah
        retlw   b'00001101'  ;  4    0D
        retlw   b'00000101'  ;  5    05
        retlw   b'10000101'  ;  6    85
        retlw   b'11000101'  ;  7    C5
        retlw   b'11100101'  ;  8    E5
        retlw   b'11110101'  ;  9    F5
        retlw   b'00000000'  ;  :    00
        retlw   b'00000000'  ;  ;    00
```

**Table 6.8 (continued)**

```
retlw   b'00000000'   ;   <   00
retlw   b'10001101'   ;   =   8D
retlw   b'00000000'   ;   >   00
retlw   b'01100110'   ;   ?   66
retlw   b'00000000'   ;   @   00
retlw   b'01000010'   ;   A   42
retlw   b'10000100'   ;   B   84
retlw   b'10100100'   ;   C   A4
retlw   b'10000011'   ;   D   83
retlw   b'00000001'   ;   E   01
retlw   b'00100100'   ;   F   24
retlw   b'11000011'   ;   G   C3
retlw   b'00000100'   ;   H   04
retlw   b'00000010'   ;   I   02
retlw   b'01110100'   ;   J   74
retlw   b'10100011'   ;   K   A3
retlw   b'01000100'   ;   L   44
retlw   b'11000010'   ;   M   C2
retlw   b'10000010'   ;   N   82
retlw   b'11100011'   ;   O   E3
retlw   b'01100100'   ;   P   64
retlw   b'11010100'   ;   Q   D4
retlw   b'01000011'   ;   R   43
retlw   b'00000011'   ;   S   03
retlw   b'10000001'   ;   T   81
retlw   b'00100011'   ;   U   23
retlw   b'00010100'   ;   V   14
retlw   b'01100011'   ;   W   63
retlw   b'10010100'   ;   X   94
retlw   b'10110100'   ;   Y   B4
retlw   b'11000100'   ;   Z   C4
```

## Formatting for transmission

In this telemetry module, the calculated values are sent back to the user as CW messages. The data in the three packed BCD bytes, up to five digits from 0 to 9, have to be coded into a CW message, with the decimal point inserted in the correct place. As full five-digit accuracy is not necessary for any of the measurements needed here, only the most significant digits are extracted for sending back. We met CW message generation in Chapter 4 where a beacon keyer sent pre-programmed characters. Here the CW was stored in an inefficient way as part of the programme structure and not conducive to large, limited-memory, applications. Here a much more compact method of storing and decoding CW characters for arbitrary transmission is shown. Table 6.8 gives a listing of the routine to extract CW from the compressed data stored in `CWTable`. CW characters are stored in a compressed form in a table as one byte per character.

For each of those CW characters that consist of five or fewer symbols, such as all the letters and numbers, the three lowest bits of the stored code represent the length of the character in symbols, and the remaining five bits are coded, from left to right, such that a '0' represents a dot and a '1' represents a dash; unused positions are left at '0'. So the letter 'J' with four symbols is stored as '01110100' and '9' with five symbols as '11110101'.

The remaining punctuation characters with six symbols are treated differently. The *last* five symbols are stored as above, and if the first symbol is a dot the length code is set to 6, or '110'. If the first symbol is a dash, this becomes 7, or '111'. So now a full-stop ($\cdot - \cdot - \cdot -$) is stored as '10101110' but a comma ($- - \cdot \cdot - -$) is stored as '10011111'.

# References


[1] Data for DTMF controller chips: www.zarlink.com.
[2] Telephone line interconnection standards: www.babt.co.uk.
[3] MC3208 data sheet: www.microchip.com.

# CHAPTER 7

# More microcontroller projects

## Hand-held, low-power RF controller

In Fig 7.1 a 16F84 PIC is used as part of a low-power RF remote control. The bulk of the work within this unit is done by the remote control chips themselves which operate with a pulse-width modulated data stream. A fixed address is set by pin selection on the transmitter and receiver chips, and these addresses have to match for data to be recognised by the receiver. Four bits in the sequence are available for data, allowing up to 16 circuits to be activated. An alternative receiver chip, the HT12F, hard wires these four address bits instead and allows control of just one circuit with higher security – this is the version usually adopted for jobs such as garage door openers. Data security against false alarms is guaranteed by the operating protocol which requires the receiver to correctly decode three successive matching codes before it will trigger an output. Further reliability can be ensured by adding an RC filter to the output line, ensuring in effect that many correct codes have to be decoded before the action is triggered.

The primary use for these chips is for infra-red remote controllers as used with television sets, but the data clock rate can be set over a wide range, making them useful for carrying data on RF or by other slower media. Here, low-cost modules designed for the licence-free 432MHz band are employed. As this falls inside an amateur allocation, we do not need to comply with the limited antenna restriction for licence-free operation, and can extend the range by use of larger antennas and power amplifiers. However, as the frequency stability of these low-power modules is a few tens of kilohertz, and modulation is rather wide, there is potential for causing QRM if high radiated power is adopted.

The hand-held controller is operated from a 4.8V battery and normally the PIC is in sleep mode. It is woken up by any key being pressed, whereupon the key press is decoded and the appropriate four-bit code is sent to the HT12E. At the receiver, depending on what is wanted, either a four-to-16 line decoder chip can be used to allow all 16 circuits to be controlled independently, or just buttons corresponding to 1, 2, 4, 8 on the keypad can be activated to control each of the data lines individually without the separate decoder chip at the receiver. The PIC controls power to the transmitter module so that it is only on when a key is pressed, ensuring power

**Fig 7.1. Low-power remote control**

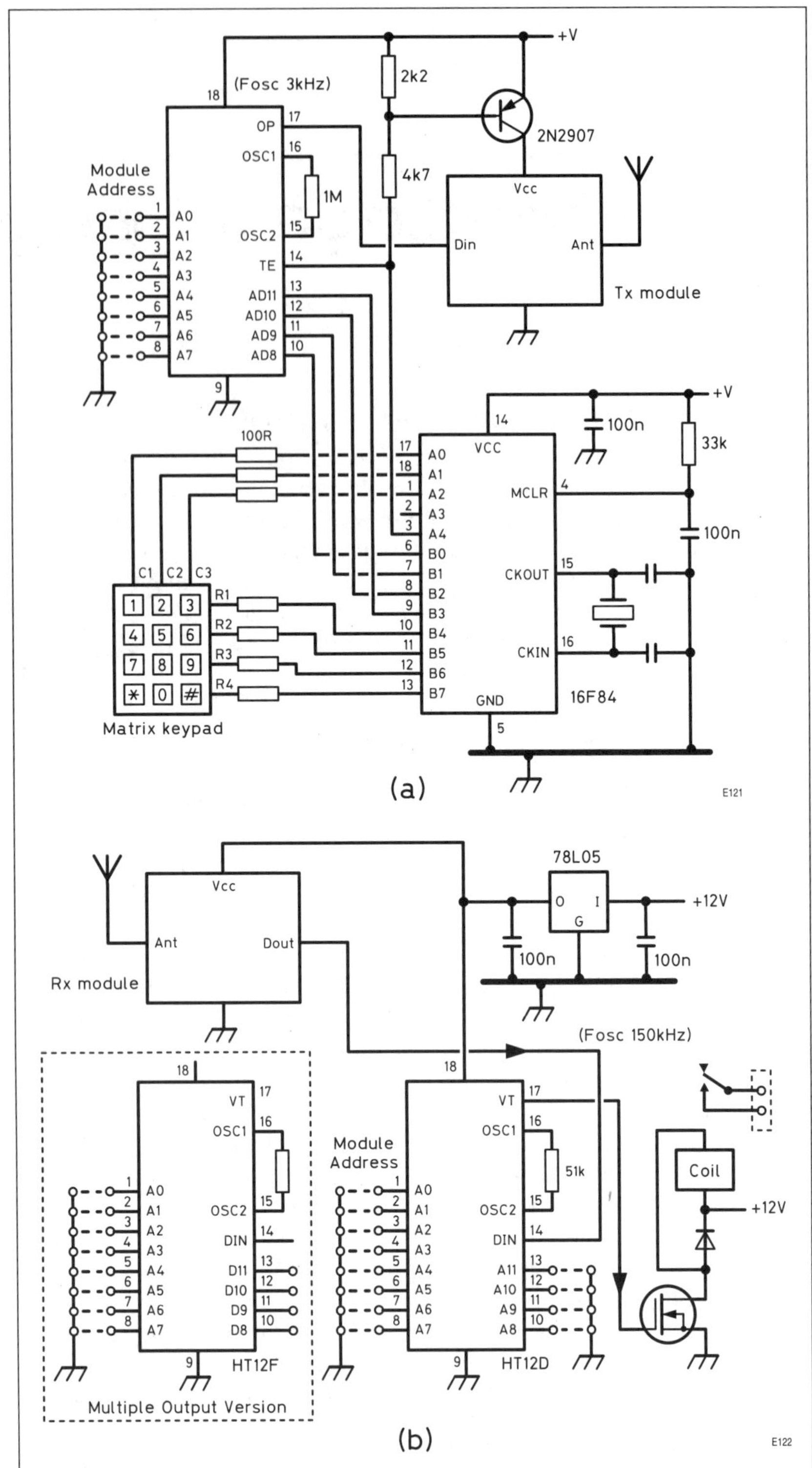

**Table 7.1. Complete PIC source code for hand-held remote controller**

```
;KEYBENC.ASM  Read hex keypad and give BCD + Strobe output
    ;timings based on 455kHz /4 clock  (8.79us)


        LIST P=16C84
        INCLUDE "P16C84.INC"
        __CONFIG b'11111111110001'  ;PU Timer on, WD Timer off, Xtal osc


        cblock      0x0D
                DelCount            ;Delay counter
                Temp
                Counter             ;General purpose counter
                KbdCode             ;
                LastKey
        endc


;code starts here


        org     0
boot    nop
        clrw
        movwf   INTCON              ;disable interrupts
        goto    startup             ;jump to main code
ints    retfie                      ;interrupt service routine code

startup
        bsf     STATUS,RP0          ;ram page 1
        movlw   b'00000000'         ;Set A as outputs  Column address + nstrobe
        movwf   PORTA               ;
        movlw   b'11110000'         ;Port B  Row inputs and data output
        movwf   PORTB
        bcf     OPTION_REG , NOT_RBPU    ;enable pull ups
        bcf     STATUS,RP0          ;ram page 0
        clrf    KbdCode
        bsf     INTCON , RBIE       ;PortB change interrupt enable

MainLoop                            ;loop here forces strobe low
        movf    PORTA , W
        iorlw   b'00001111'
        movwf   PORTA               ;preserves Strobe, columns all high

        bcf     PORTA , 0
        movlw   0x00
        movwf   KbdCode
        call    Delay
        comf    PORTB , W           ;Read row code (any 0 means key pressed)
        andlw   b'11110000'
        btfss   STATUS , Z
        goto    GetRow              ;this row detected

        bsf     PORTA , 0
        bcf     PORTA , 1           ;Address col 2
        movlw   0x04
        movwf   KbdCode
```

**Table 7.1** *(continued)*

```
        call    Delay
        comf    PORTB , W
        andlw   b'11110000'
        btfss   STATUS , Z
        goto    GetRow

        bsf     PORTA , 1
        bcf     PORTA , 2        ;Address col 3
        movlw   0x08
        movwf   KbdCode
        call    Delay
        comf    PORTB , W
        andlw   b'11110000'
        btfss   STATUS , Z
        goto    GetRow

        bsf     PORTA , 2
        bcf     PORTA , 3        ;Address col 4
        movlw   0x0C
        movwf   KbdCode
        call    Delay
        comf    PORTB , W
        andlw   b'11110000'
        btfss   STATUS , Z
        goto    GetRow
        bsf     PORTA , 4        ;Kill Strobe as no key pressed

        movlw   b'00010000'      ;Set columns all low so any key press
        movwf   PORTA            ;   will awaken from sleep
        bcf     INTCON , RBIF    ;Clear port B change flag
        sleep
        nop                      ;Wake up to here if key pressed
        goto    MainLoop
;------
GetRow                           ;Enter with Bits 2/3 of KbdCode showing Col count
        call    Delay    ;10ms Debounce delay
        call    Delay
        call    Delay
        call    Delay
        call    Delay
        call    Delay
        call    Delay
        call    Delay
        call    Delay
        call    Delay

        btfsc   PORTB , 4
        goto    Row2
        movlw   0
        addwf   KbdCode
        goto    RowOk
Row2
        btfsc   PORTB , 5
        goto    Row3
```

**Table 7.1 (continued)**

```
          movlw   1
          addwf   KbdCode
          goto    RowOk
Row3
          btfsc   PORTB , 6
          goto    Row4
          movlw   2
          addwf   KbdCode
          goto    RowOk
Row4                            ;must therefore be column 4
          movlw   3
          addwf   KbdCode
RowOk
          movf    KbdCode , W
          subwf   LastKey , W
          btfsc   STATUS , Z    ;if last two reading are equal, valid press
          goto    KeyOk
          movf    KbdCode , W
          movwf   LastKey       ;save key for check next time
          goto    MainLoop
KeyOk                           ;KbdCode contains a valid col/row number
          movf    KbdCode , W
          call    Table
          movwf   Temp
          bcf     STATUS , C

          movf    Temp , W      ;Put BCD Output to 4 LSBs of PORTB
          movwf   PORTB
          bcf     PORTA , 4     ;Strobe after valid check
          call    LongDelay     ;inhibit another key for 500ms
          goto    MainLoop
;. . . . . . . . . . . . . .
Delay
          movlw   d'21'         ;delay = 5N + 9  = 1ms
          movwf   DelCount
DelLoop
          nop
          nop
          decfsz  DelCount
          goto    DelLoop
          nop
          nop
          nop
          nop
          return
;. . . . . . . . . . . . . .
LongDelay
          movlw   d'250'
          movwf   Counter
LongDelLoop
          call    Delay         ;500 calls to delay
          call    Delay
          decfsz  Counter
          goto    LongDelLoop
```

**Table 7.1 *(continued)***

```
        return
;...............
Table                   ;3x4 keyboard,   KdbCode format   0000CCRR
        addwf   PCL             ;Valid for codes 0 to 254
        retlw   1               ;Row 1 , Col 1
        retlw   4
        retlw   7               ;table for 3x4 keypad
        retlw   0x0A            ;*
        retlw   2
        retlw   5
        retlw   8
        retlw   0               ;
        retlw   3
        retlw   6
        retlw   9
        retlw   0x0B            ;#
        retlw   0x0C            ;  Col 4 not always implemented
        retlw   0x0D            ;
        retlw   0x0E            ;
Tend    retlw   0x0F            ;

        if (( Table & 0xFF) >= (Tend & 0xFF))
            MESSG "Warning,  table crosses page boundary"
        endif

        if (Table >= 0x100)
            MESSG "WARNING Table start above 0x100, ensure PCLATH is set"
        endif

;...............
        end
```

consumption is kept low enough to give hundreds of hours operation from the battery in the hand-held unit.

## Filter selection

This unit is designed to automatically switch relays that select the low-pass filters at the output of an HF PA. Most HF power amplifiers are broad-band devices, covering at least 2 to 30MHz, but harmonic filters have to be selected depending on the frequency of operation – usually there are five or six low-pass filter elements with sub-octave frequency cut-offs. If the PA is designed to be no-tune in its operation then it is a pity to spoil this 'transparency' by having to manually select the low-pass filter to use. The solution usually adopted is to let the transceiver 'tell' the PA which band it is on by sending a tuning voltage, but why not just measure the transmitted frequency and set the filters accordingly?

The circuit is shown in Fig 7.2 and works as follows. A portion of the RF signal is converted to a logic level and its frequency is divided by four in a pair of flip-flops. Then it is applied to the PIC counter-timer input. The PIC

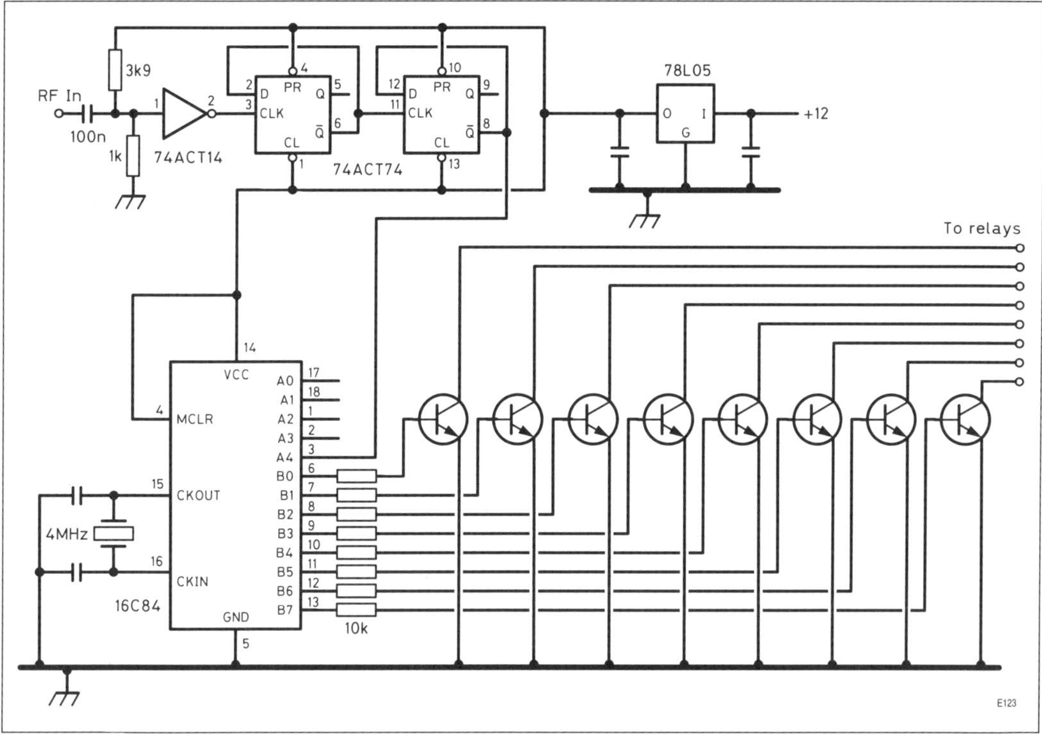

Fig 7.2. HF PA automatic relay selector circuit diagram

timer is configured to rapidly measure the frequency and then, by making use of a small look-up table, choose which one of the output ports to activate, which in turn switches into circuit the appropriate low-pass filer. In fact the external prescaler is not really necessary as the 16F84 internal prescaler will operate with up to 50MHz input; this snippet of useful information was not known at the time of designing the project, hence the pair of flip-flops!

With typically six filters that can be switched, it is not necessary to measure the frequency very accurately – all that is really required is to be able to measure the frequency to something like the nearest megahertz. With an eight-bit counter in the PIC, an upper operating frequency limit of 30MHz means a frequency resolution on $30/256 = 117$kHz can be achieved. Here a gate time of 250µs is employed, coupled with a total of 32 times prescaling of the RF. This frequency division is made up from four in the external divider and another divide by eight within the PIC. So the actual frequency measurement resolution is $1/250$µs $\times 32 = 128$kHz. The complete software listing is given in Table 7.2. The frequency is continuously measured in the MainLoop part of the programme, and converted to a frequency 'number' based on the boundary values defined at the beginning of the programme. Here only six possible values are used, hence the five boundary values defined as F1 to F5. The F0 value is a frequency threshold below which any measurement is deemed to be 'no signal present'. Every time a frequency number is obtained, it is compared with the previous two readings, and

**Table 7.2. Complete software listing for automatic HF PA filter switching**

```
;    FILTSW1A.ASM  Measures Tx frequency /4  via timer, based on 4 MHz PIC
;    clock and sends  code to Port B for driving normal relays
;    Bit 0 - lowest frequency etc.


        LIST  P=16F84
        INCLUDE  "P16F84.INC"
        LastSet         equ     0x0D    ;Holds current freq code/relay state
        ThisFreq        equ     0x0E    ;Temp latest frequency measurement
        Temp            equ     0X0F
        PortTemp        equ     0x10    ;Temp working register for Port B
        DelReg          equ     0x11
        Freq            equ     0x12    ;Current frequency code measurement
        DelReg2         equ     0x13    ;needed for relay pulse delay
;Filter lower boundaries (units of 128 kHz, max freq > 32.6 MHz overflows)
        F0              equ     d'11'   ;1.41 MHz Test frequency for no signal
        F1              equ     d'17'   ;2.18 MHz
        F2              equ     d'31'   ;3.97 MHz
        F3              equ     d'58'   ;7.42 MHz
        F4              equ     d'117'  ;14.98 MHz
        F5              equ     d'172'  ;22.02 MHz
        GateConst       equ     d'49'   ;delay constant for counter gate = 250us


        org     0
        nop
        clrw
        movwf   INTCON          ;disable interrupts
        goto    startup         ;jump to main code
ints    retfie                  ;interrupt service routine code


startup
        clrwdt
        bsf     STATUS,RP0      ;ram page 1
        movlw   b'11100010'     ;External input, Prescalar /8
        movwf   OPTION_REG
        movlw   b'00010000'     ;
        movwf   PORTA           ;A4 as input for Fin, rest as outputs
        movlw   b'00000000'     ;Port B as outputs for relay drivers
        movwf   PORTB
        bcf     STATUS,RP0      ;ram page 0
        movlw   b'00100000'
        movwf   PORTB           ;start with highest freq as default

MainLoop                        ;Frequency CODE used throughout
        call    GetFreq         ;Frequencies coded as 0 - 6 represented
        movwf   Freq            ;as boundaries for relays
        btfsc   STATUS , Z      ;Check for no signal
        goto    MainLoop


        call    GetFreq         ;measure again
        subwf   Freq , w
        btfss   STATUS , Z      ;if not the same forget it
        goto    MainLoop


        call    GetFreq         ;Three identical measurements, its probably
```

**Table 7.2 (*continued*)**

```
        subwf   Freq , w            ;the correct value by now
        btfss   STATUS , Z
        goto    MainLoop

        movf    Freq , w
        subwf   LastSet , w         ;Has it changed from last time ?
        btfsc   STATUS , Z
        goto    MainLoop
        movf    Freq , w
        movwf   LastSet             ;Update the current frequency

        movlw   1
        movwf   PortTemp
        bcf     STATUS , C          ;make sure 1 doesn't get shifted in
PortLoop
        decfsz  Freq                ;lowest freq is 00000001
        goto    Shift
        movf    PortTemp , w
        movwf   PORTB
        goto    MainLoop
Shift
        rlf     PortTemp            ;move 1 to left, Freq number of times
        goto    PortLoop
;----------
GetFreq
        clrf    ThisFreq            ;Local variable for frequency code
        clrf    TMR0
        movlw   GateConst
        movwf   DelReg
        nop
        nop
        nop
GateLoop                            ;delay between TMR set and read =
        nop                         ;5N + 5
        nop
        decfsz  DelReg
        goto    GateLoop
        movf    TMR0 , w

        movwf   Temp                ;
        sublw   F0                  ;
        btfss   STATUS , C
        incf    ThisFreq

        movf    Temp , w            ;Temp contains gated count
        sublw   F1                  ;test for >F1,  W = F1 - W, If -ve then C=0
        btfss   STATUS , C          ;means greater than F1, so inc. Freq Code
        incf    ThisFreq

        movf    Temp , w            ;each time > threshold, incr ThisFreq value
        sublw   F2
        btfss   STATUS , C
        incf    ThisFreq
        movf    Temp , w
```

**Table 7.2 *(continued)***

```
        sublw   F3
        btfss   STATUS , C
        incf    ThisFreq
        movf    Temp , w

        sublw   F4
        btfss   STATUS , C
        incf    ThisFreq
        movf    Temp , w

        sublw   F5
        btfss   STATUS , C
        incf    ThisFreq

        movf    ThisFreq , w
        return                  ;Returns frequency code in W
;––––––––––
        end
```

only if three consecutive identical readings are obtained is it considered to be a perfect result. Making use of three readings in this way ensures there will be no spurious switching if a frequency measurement error occurs, due to a low RF level for example. The valid frequency number is then compared with the current value and, if these are different, the new code is sent to the relay driver port. Since there is no shortage of ports available on the PIC for this project a serial driver chip such as the UCN5841 is not necessary and cost is saved by using simple transistor drivers for the relay interfacing.

## PicATUne automatic antenna tuner

The PicATUne, designed by Peter Rhodes, G3XJP, was originally described in *RadCom* [1], and contains some novel ideas to simplify design and reduce costs. Here we limit the description to a brief outline of the tuning algorithm and PIC software. Fig 7.3 shows a block diagram of the PicATUne concept from which it can be seen that the critical components are a SWR detector head and a bank of relay switched inductors and capacitors making up the antenna tuning assembly. Fig 7.4 shows the complete circuit diagram of the logic board. The PIC first measures the frequency and looks in memory to see if a match for this frequency already exists. If it does, these values are set and tested for a low SWR reading. If no values exist, or the SWR is too high due to a changed antenna, for example, the PIC software attempts to find a new match. It does this by measuring the reflected signal level, presented as a DC voltage from the SWR detector head, and then making a change to the L or C values. The change in reflected signal is measured again to determine if the result is better or worse than before. The PIC software is thus simulating an operator adjusting a manual ATU by reference to a SWR bridge. A more comprehensive description of the tuning algorithm is given in reference [1].

Once a match has been obtained the frequency is measured by the PIC counter/timer and the L/C settings are stored in non-volatile EEPROM memory for future access. A novel trick adopted in this design is the measurement of an analogue voltage without the cost of including an A/D converter chip. The RA3 pin is normally set as an input. When a reflected power measurement is required, RA3 is changed to an output pin, pulsed high and immediately reconfigured as an input pin. The time taken for RA3 to fall below the '1' threshold is then determined, giving a measure of the voltage across C48. The width of the high pulse is adjusted with the cal relay energised – and therefore no reflected power – to calibrate the zero point. The process is crude but effective, given only that the requirement is to determine the direction of change as the L and C values are altered.

A serial input shift register allows the large number of relay drive lines to be controlled by making use of just three PIC outputs, and separate driver transistors are used instead of a dedicated serial relay driver as this actually saves PC board space.

The final part of note is the frequency measurement routine. The system described in the previous section for filter selection is not adequate here as a higher frequency measurement resolution is needed and a clever workaround is employed to enable the contents of the internal prescaler, set to divide by its maximum possible of 256, to be read. Normally these lowest eight bits of the count, which cannot be read out in software, would be lost in the frequency measurement process, reducing accuracy to below that needed. To count RF cycles both RB4 and RA4 are set as inputs for the counting gate time – in this case 400µs. At the end of the gate period, RB4 is changed to an output pin to freeze the count – this is then toggled high/low until the main counter register changes, at which point the prescaler finally overflows. Knowing

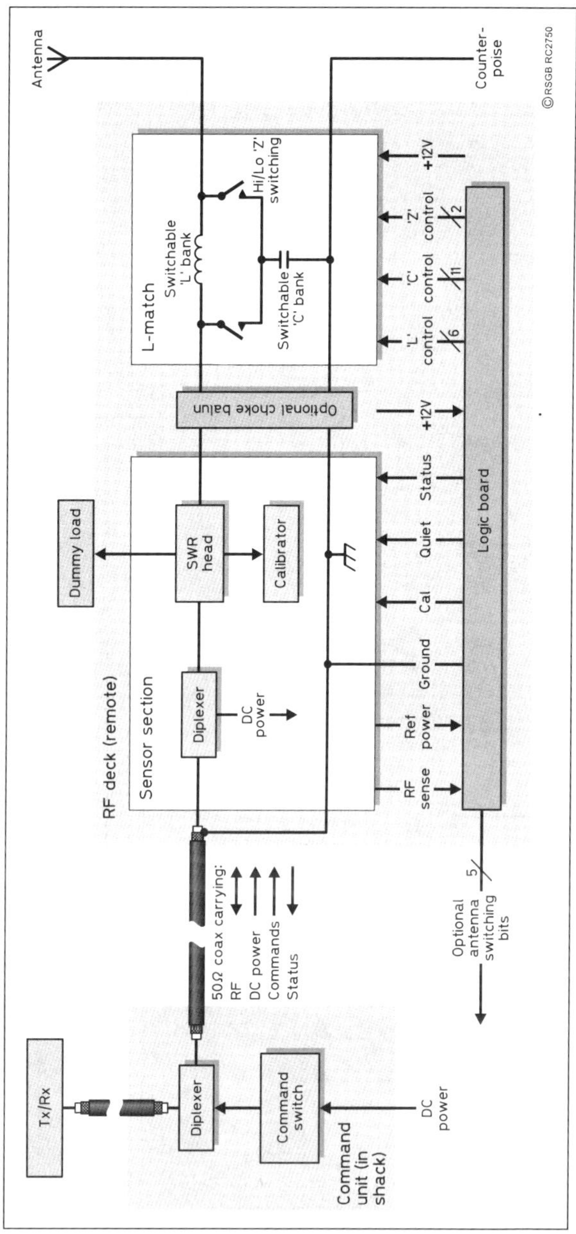


**Fig 7.3. PicATUne block diagram**

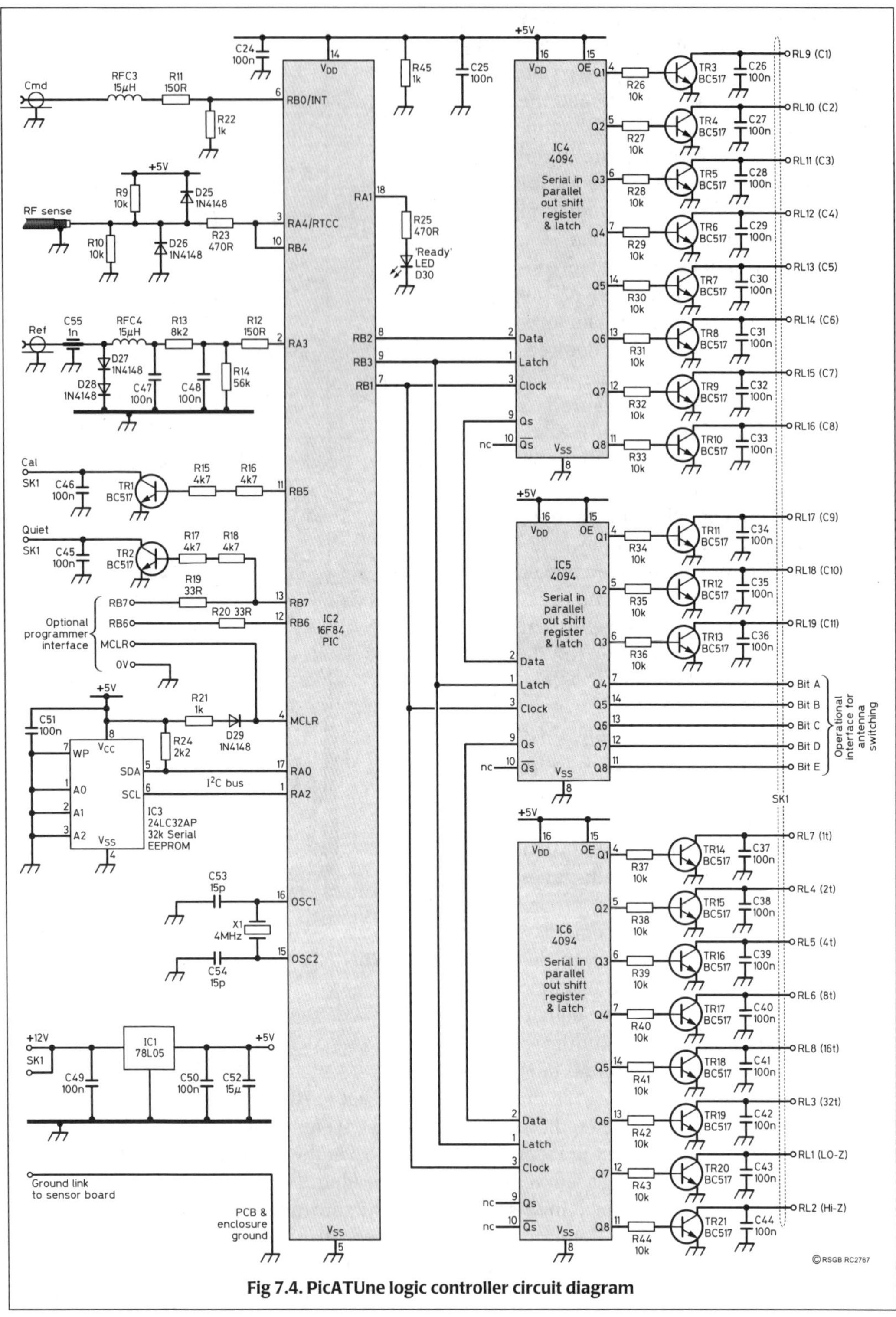

**Fig 7.4. PicATUne logic controller circuit diagram**

how many times RB4 was toggled, the value of the eight most-significant bits in the main counter, as well as the eight least-significant bits in its prescaler are determined, giving a full 16-bit count subject to an error of $\pm 1$ count, or 2.5kHz.

## GPS-locked frequency stabiliser

This low-cost, GPS-locked frequency source is designed specifically for low-data-rate signalling on the LF bands. Phase noise and short-term frequency stability preclude its general use at HF or above; for these frequencies other more conventional frequency standards are preferable, based around crystal oscillators of inherently higher stability.

A conventional synthesiser or phase-locked loop (PLL) approach is impractical for locking a frequency source to the 1 pulse per second (pps) signal from a GPS receiver. The only reference frequency is at 1Hz so a PLL would need to have a bandwidth considerably less than this, meaning that the loop would only lock up after many tens of minutes or hours. Furthermore, the stability of the basic oscillator element – the voltage-controlled oscillator or (VCO) – would have to be such that it did not drift by more than 1Hz during a time comparable with the loop bandwidth, otherwise lock could never be achieved. A high-stability VCO is needed and, for a design operating at a few megahertz, the required stability would be less than 1 part per million (ppm). Voltage-controlled crystal oscillators (VCXOs) to this accuracy, also usually temperature controlled, are available but are expensive and only occasionally appear on the second-hand surplus market as part of test equipment. Brooks Sheera, W5OJM, has designed an excellent GPS-locked high-stability frequency source using such an oscillator, which was described in *QST* July 1998. This does require several hours to achieve lock-up and is the sort of precision test equipment that should be left running continuously and not be turned off every day.

However, where short-term stability is of less importance, another technique can be used. The source described here is based roughly on the old Huff and Puff stabiliser, first described by Klaus Spaargen, PA0KSB, in 1973. This design stabilised a 'good' LC-tuned VCO to give it crystal oscillator stability, in steps of 10Hz. In essence, the design used a 1-bit frequency counter – a single flip-flop clocked by pulses divided down to 10Hz from a crystal oscillator. Depending on whether the VCO frequency was high or low of the appropriate 10Hz step, the output of the flip-flop ends up at a 1 or 0 and is applied via a low-pass filter to a variable-capacitance tuning diode on the LC oscillator, which is ramped up or down to maintain the long-term frequency. The output would then be continuously hunting just above or just below the nearest 10Hz step.

The problem with this design was that if the LC oscillator drifted by more than one step, eg 10Hz, the locked frequency would therefore slowly move in jumps of 10Hz. There were numerous improvements to the original design in the following years, several adding more bits and resolution to the frequency counter, allowing a higher inherent drift, but the Huff and

Puff gradually died out as frequency synthesisers took over. Synthesisers all rely on phase locking a VCO to a master reference, usually at a reference frequency of a few kilohertz, so consequently the PLL can lock up quickly and short-term stability is not an issue – long-term stability is then purely a function of the reference which can be made as good as is wanted.

## Circuit description

This design applies the Huff and Puff technique to a simple VCXO without a crystal oven to maintain a source frequency that hunts either side of a specified 1Hz multiple, but uses an extended eight-bit frequency counter to avoid the possibility of the locked frequency drifting in jumps of 1Hz. The VCXO operates at 4.194304MHz ($2^{22}$Hz) for reasons that will be described later, but any frequency that is a multiple of 1Hz may be employed. It is also easier to understand the concept if a frequency that is an exact multiple of 256Hz is initially chosen. Refer to the circuit diagram, Fig 7.5.

A straightforward crystal oscillator is built with CMOS gates, with a variable-capacitance diode connected across the crystal to shift its frequency either side of nominal by a few tens of hertz. Choose the diode and coupling capacitors to ensure that it is not possible to pull the frequency more than 128Hz over the full input voltage swing of 0–5V. This oscillator drives a synchronous eight-bit counter made from a pair of 74HC161 chips, the eight outputs from which are connected to a 74HC374, eight-bit D-type latch. This is triggered by the rising edge of the 1 pps signal from a GPS receiver so that each second the output of the latch contains the lowest significant eight bits of the count. This counter overflows many times for each counting interval, but for any frequency that is an *exact* multiple of 256Hz, the counter will overflow to the same point each time, and the count latched by each seconds pulse will then be a constant number. If the frequency is not a multiple of 256Hz, but is still an exact 1Hz multiple, the count will change from one second to the next in a predictable manner. For example, a 5MHz signal would give successive counts (assuming a start at zero) of 00, 64, 128, 192 before the sequence repeats, which is completely predictable. However, for this description we will continue with a frequency that is a multiple of 256Hz so the ideal count stays a constant. If the frequency departs slightly from its correct value, the residual count will steadily increase or decrease by the magnitude of the frequency error and, if not corrected, the counter will eventually wrap round.

What we now do is to read the count every second, then drive a charge pump circuit in a direction such as to correct the frequency. Software in a 16F84 PIC processor, using the trigger pulse from the GPS, reads the latched counter value and calculates an error value from the expected count every second. For multiples of 256Hz this merely involves subtracting a constant, which can conveniently be a value of 128 – half the counter length. If other frequencies are wanted, the 'correct' value will increment by the frequency modulo 256, every second. The sign and magnitude of the error value obtained is then a measure of the phase, and hence frequency, error and is used as follows.

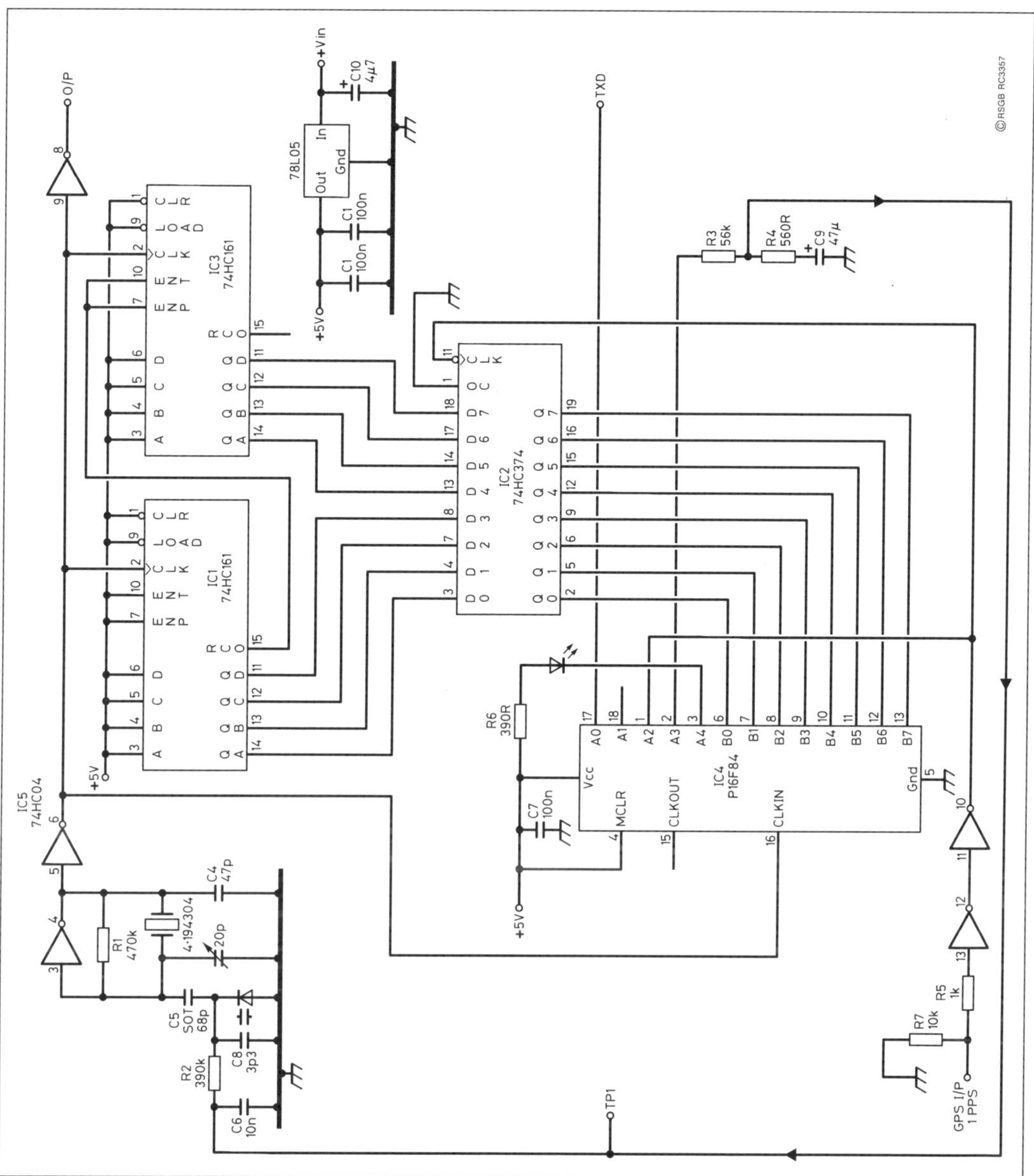

Fig 7.5. GPS-locked frequency stabiliser circuit diagram

Port A3 on the PIC is normally maintained at a high impedance by setting it as an input. Every second, except for the single case where the error value is zero, this port is set as an output and strobed high or low, depending on the sign of the frequency/phase error. The duration of the strobe pulse is proportional to the magnitude of the error so that an error count of 128 leads to a strobe pulse of 0.64 seconds, either high or low, down to the minimum resolution of 5ms for an error count of one. This strobe pulse is then taken via an RC network to drive the variable-capacitance diode on the VCXO. The RC constants were determined by a combination of experience, trial and error, serendipity and luck to give an acceptable compromise between lock-up time, chirp and pull-in range. To assist lock-up, after turn-on the PIC software applies a precise square wave to the charge pump for 20 seconds, to force the capacitor to mid-voltage and near to the nominal operating frequency. The loop then only has to make minor corrections to get the correct frequency and phase (the counter mid-point). The use of a three-state, high-impedance port which is only briefly pulled high or low considerably reduces the residual chirp over that of a continuously changing low-impedance connection, as in the original PA0KSB concept.

As an additional aid to debugging, the error value is output from the PIC as an RS232 signal at 1200 baud. The format is a decimal number ranging from −128 to 127, followed by a space, for display on a standard ASCII terminal.

## Performance

With the RC values and variable-capacitance diode specified, after the 20s initialisation period the loop was fully stabilised after running for about two minutes. The frequency as measured on a counter was exact, when using a 10s counter gating time. By monitoring the output on a vectorscope referenced to a high-stability source, every second, on average, the phase would rapidly rotate by a value between 180 to 500° (0.5 to 1.5 cycles) then return more slowly. This appeared as a slight audible chirp of a few hertz. However, the mean phase stays constant, ie the vectorscope always averages out the clockwise and anti-clockwise rotations to zero. The error value transmitted from the RS232 port usually stabilised at a small positive number rather than zero. The reason for this is not clear, but may be due to leakage around the charge pump circuitry, or rectification of the RF in the diode, causing a constant offset. Its effect does not appear to be important to operation. Without programming in the 20s initialisation period, there were times, depending on the initial random starting count, when the loop would fail to lock up at all, and other times when lock would occur after a few minutes.

By using the output as a clock for an AD9850 DDS chip, the output frequency is divided down and the phase blip is reduced by the same proportion. At 137kHz, for example, the 360° blip at 4MHz appears as a blip of 11° every second, always returning to the same value. When averaged out over a 30s signalling element the net phase shift is very close to zero.

A LED flashes with the 1 pps signal, and the duration of flash is related to the width of the charge pump pulse. Therefore, when locked the LED gives a short flash but, during the lock-up phase, the flash is of varying duration and, if the GPS pulse is not present, the LED does not flash at all.

## Construction and setting up

The circuit layout is not critical, apart from that around the VCXO. Here, wires need to be as short as possible and the loop filter components need to be mounted close to the variable-capacitance diode to avoid picking up interference which is then coupled onto the output signal. A double-sided PCB is illustrated which makes use of normal 2.54mm pitch DIL ICs, but surface mount resistors and capacitors. The layout is given in Fig 7.6.

There is a certain amount of initialisation of the circuit necessary before satisfactory operation can be guaranteed. Firstly, the VCXO has to be set to the correct frequency. Before soldering in the RC loop filter components, connect a potentiometer to the variable-capacitance diode decoupling resistor, shown as TP1 on the diagram. Set to mid-rail at 2.5V and adjust the preset capacitor for a frequency as close as possible to the correct value. Swing the tuning voltage over the range 0–5V and ensure that the output frequency shifts by no more than about 100Hz in either direction. Much more than this, and a false lock 256Hz away is theoretically possible if the crystal drifts over time; much less that this, lock-up time and drift may be a problem. To change the tuning range, alter the value of the Select On Test capacitor experimentally. Once the VCXO tuning has been set up, insert the remaining loop filter components. Connect the 1 pps output from a GPS receiver to the board and switch on. It may be convenient to monitor the loop tuning voltage with a high-impedance meter at this stage. Connect PIC port A0 to the RXD line of an ASCII terminal set to 1200 baud, eight bit, no parity to monitor the loop locking progress.

For the first 20s after switch-on, the LED will stay off and the voltage at TP1 should rise to 2.5V, at which time the output frequency will be close to

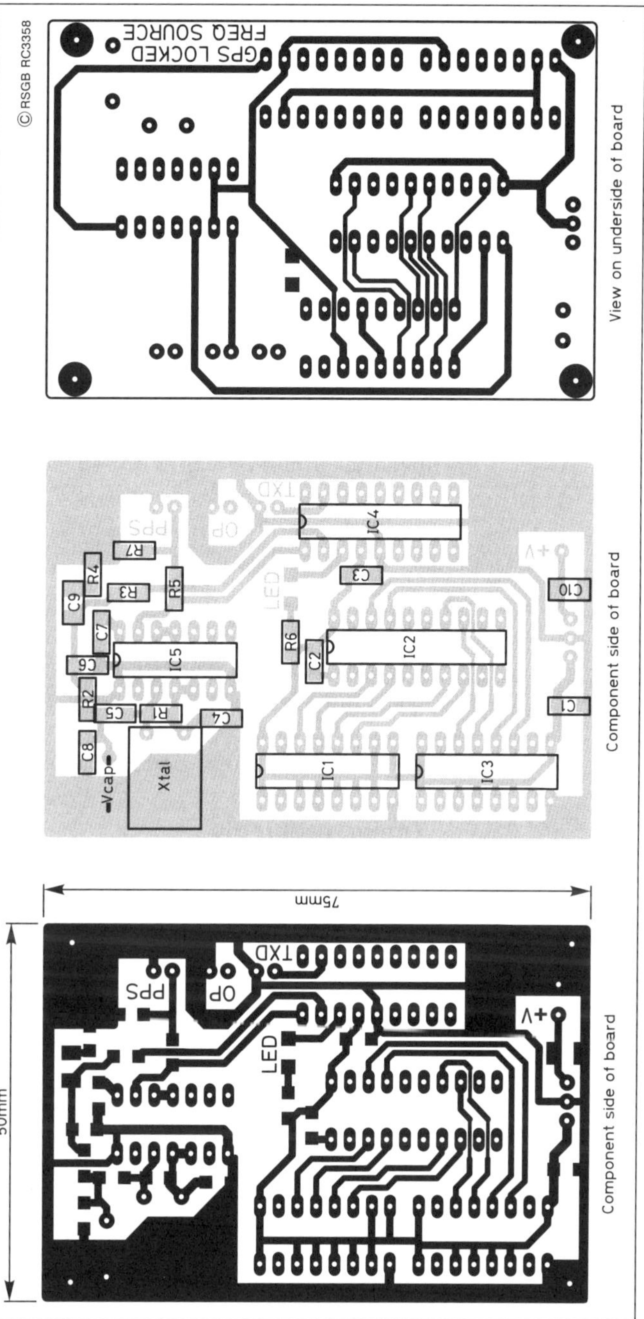


Fig 7.6. GPS locked frequency source PCB layout

the correct value if the setting-up procedure was followed carefully. After this time has elapsed, the LED will start to flash in synchronism with the GPS pulses, the flash duration appearing to vary in an apparently random manner. If the counter error value is being monitored, it will also show a rapidly varying number each second but, after about 30–60s, a pattern will start to emerge and the value will gradually converge to a constant value, not far removed from zero; the duration of the LED flashes will also shorten. The voltage at TP1 should now be stable, and after a few minutes the loop should have stabilised, at which point the output frequency is locked. If you listen to this signal on an SSB receiver, it should give a slight blip every second which should be just about audible, but how much depends on how musical your ears are!

There is plenty of scope for further experimentation, particularly around the loop, to decrease chirp/phase blips and improve lock-up times. One idea that would be worth investigating would be to have a non-linear relationship between error value and pulse width. This is something that is straightforward to implement in software but complex in a PLL built completely in hardware.

## PIC software

The PIC software STABIL01 is available from the RSGB website under that name. An enhanced version STABIL02 is also available with several minor improvements. Detailed operation of the software can be seen by examining the source code in the .ASM files.

## Known problems

1. The loop error value is rarely stabilised at exactly zero during operation. This is probably caused by rectification of the RF in the VCXO by the variable-capacitance diode, which the loop has to fight. It could be cured by adding an op-amp buffer to isolate the diode drive from the loop filter itself, but the effect does not appear to cause a problem so no buffer has been included. Increasing the value of the 390kΩ resistor feeding the diode reduces this effect, but this shouldn't be made too high as spurious pick-up could then become more of a problem. Charge pump capacitor leakage will show the same effect if any is present, and is particularly noticeable at any time a voltmeter is connected to TP1.
2. The latching of the counter by the GPS pulse is asynchronous to the counter increment, so there is a probability (a certainty in fact) that at some point the reading will try to be latched at the very instant it is changing; minimising this, in fact, was the reason for using a synchronous counter here. The only way to prevent the situation from occurring at all would be to add some gating or second-level buffering to the latch, which would be horribly complicated for a simple design such as this. Fortunately there is a simple work-around. During testing it was observed that a read like this always appeared to result in a latched value of all ones. The software detects this condition and ignores that particular value,

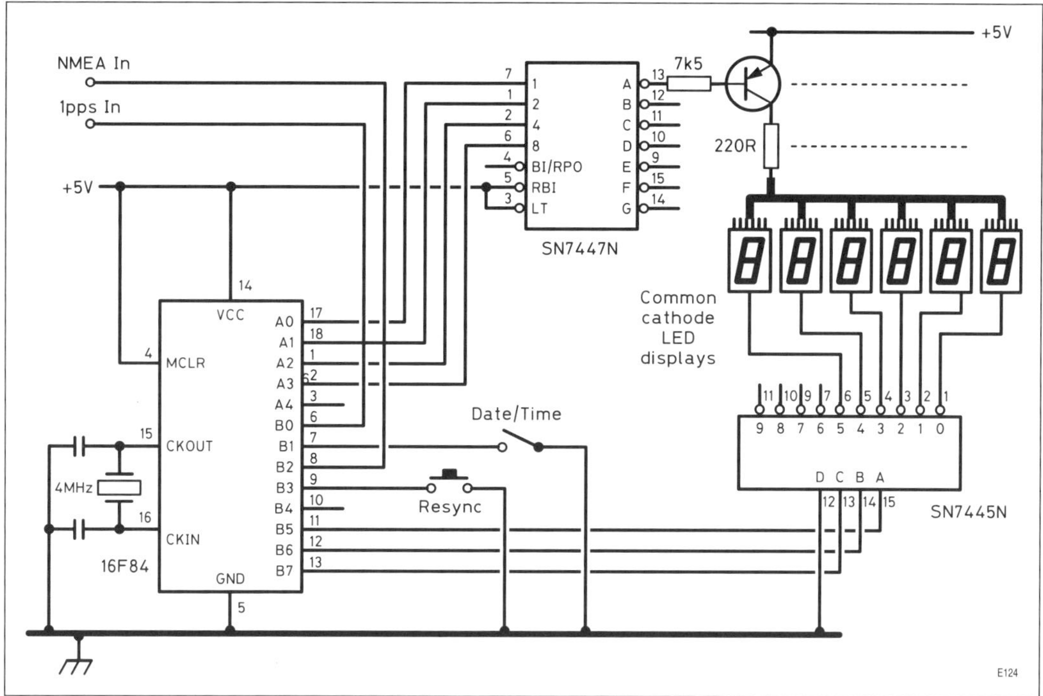


Fig 7.7. GPS clock circuit diagram

treating it as if a value of 128 had been read instead and issuing no pulse. The loop can never be so stable that this glitch situation will occur repeatedly, and the tracking continues uninterrupted at the next pulse. The resultant 2s delay may cause a slightly larger blip than usual, but this should not be unacceptable in practice

## Automatically set GPS clock

This PIC-based design also interfaces to a GPS module and makes use of the 1 pps signal and the serial NMEA output, using the data contained in the NMEA message to automatically set a clock. The NMEA sentence structure was illustrated in Chapter 3, and the GPRMC sentence is used here as it contains both time and date information. A display of either data or time is made on a six-digit LED display which is operated in a multiplexed design as described in Chapter 4.

The circuit diagram is shown in Fig 7.7 which shows how separate digit-driver and seven-segment decoder-driver chips are used to free up sufficient PIC I/O lines. The NMEA output from the GPS module is connected directly to the PIC; the Garmin GPS25 module gives simultaneous outputs of both TTL and RS232 level signals. We make use of the former here, so the polarity of the RS232 interface is opposite from that in most of the designs in this book and can be connected directly to the PIC input. Some GPS modules only support true RS232 levels and these will require a series resistor of around 5.6kΩ to avoid overdriving the input port, with a

**143**

modification to the software to invert the polarity. Alternatively a separate single transistor or logic gate inverter could be included.

The complete software is contained in the file GPSCLOCK available from reference [1] but a basic description follows. During normal operation the software sits in a loop, multiplexing data stored in the hours/minutes and seconds registers to each of the six digits in the display in turn. During this period it is also monitoring the user switches for display of date rather than time, or, if requested, re-syncing to the NMEA sequence. This is rarely if ever necessary but has been included 'just in case'. The GPS 1 pps input causes the software to jump to the interrupt routine which progressively increments the seconds counter then tests for rollover to the next minute, hour etc right the way through to the year storage register. Leap year correction is included using the four-year rule, but no more – it will fail in the years 2100, 2200 and 2300 but is unlikely to be a problem, and can be solved just by pressing the re-sync button which will allow correct operation for another hundred years. This piece of software was written during the last months of 1999 and tested when it successfully showed the exact start of the new millennium.

At switch on, the 1 pps update interrupt is disabled and the software has to get the correct time from the NMEA message. It does this by receiving all characters appearing in the NMEA message and pushing them onto a stack, ensuring the last six are always available. The stack is continuously checked for the header '$GPRMC' and, when this text appears, the programme starts looking for the commas that delimit the message parameters. The commas are counted and as soon as this count reaches two, we know that the data just preceding it was the time of the last pulse in HHMMSS format, and that this information was saved to a known position on the stack, ready to be transferred into the time-holding registers. Receipt of the 10th comma indicates that the date, also six digits, has just been sent and this can be transferred similarly. Once the correct date and time have been read into the counter registers, a flag is set to indicate data validity, the PPS interrupt is enabled, and software control proceeds to the main display multiplexing loop with the interrupt routine updating the date and time. By testing for other comma counts the position of the observer could be extracted, but the stack size would then have to be increased and the display altered to accept the increased number of digits required.

## Converting code for the 16F628

Many of the PIC programmes described so far are targeted at the 16F84 device. This, with its Flash memory and EEPROM non-volatile data storage, make it one of the most useful chips in the family for general-purpose use, and it has become the standard 'workhorse'. However, a newer device, the 16F628, has now appeared on the scene and is set to become the new workhorse – and furthermore it is slightly cheaper than the 16F84. Its advantages over the 16F84 are greater programme memory, 2048 words instead of 1024; increased data memory with 224 bytes of internal RAM and

**Table 7.3. Example set-up code for the PIC16F628, showing some of the variations from that for the 16F84 device**

```
__config    0x3F63        ;LV Programming off , OSC EC Mode (RA6 I/O)


;INITIALISATION
    clrf    PORTA            ;Special set-up to disable comparator on 16F628
    movlw   7
    movwf   CMCON
;....................
GetEE           ;different from 16F84    EE registers all in bank 1 now
    bsf     STATUS,RP0      ;ram page 1
    movwf   EEADR
    bsf     EECON1 , RD
    movf    EEDATA , W
    bcf     STATUS , RP0    ;ram page 0
    return
```

128 bytes of EEPROM. Additional hardware includes a synchronous/asynchronous receiver transmitter (USART), extended counter timer functions and voltage comparator module. It is also possible, when using an external clock, to get access to another I/O connection on PORTA. A low-voltage programming mode is available to allow in-circuit reprogramming without having to connect a separate programmer to generate a high programming voltage – the existing 5V levels can be used.

Code written for the 16F84 will run on the 16F628, but a few changes are needed to cope with the extra peripherals and their connections to the outside world :

1. Low-voltage programming has to be set to OFF if port connection B4 is to be used – the default is ON, and the appropriate configuration bit has to be set.
2. The voltage comparator is configured as the default, and if this function is not needed, the CMCON bits must to be set to disable the comparator to gain access to PORTA.
3. The EEPROM data memory control registers are now all located in Bank1 – so existing code for accessing these has to be changed.
4. To get access to PORTA,6, set External Oscillator mode in the configuration bits.

An example of set-up code for the 16F628 is shown in Table 7.3.

## The Mini DDS (direct digital synthesis)

This project was designed by Jesper Conran as an example of what can be achieved with the high processing speed and simple instruction set available in the AVR processor. The following was extracted from his website [2].

"After looking at a design that used an Analog Devices AD9832 DDS chip, I came up with a very simple version of the DDS synthesiser, using

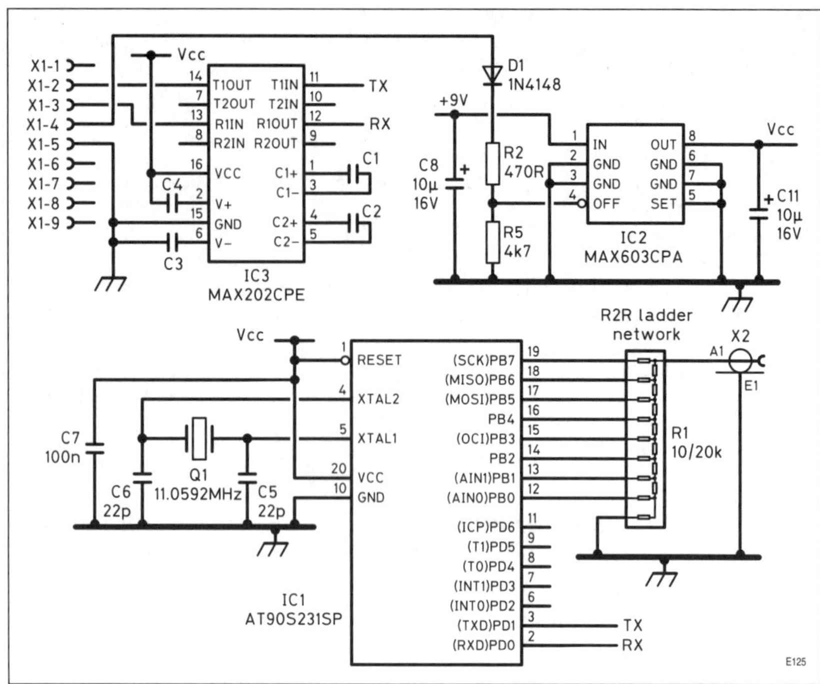

Fig 7.8. Mini DDS circuit diagram

just the 2313 and a resistor network. It's controlled over RS232 from a small Windows program, and can generate sine, sawtooth, triangle and square waves ranging from 0.07Hz to about 200–300kHz in 0.07Hz steps, depending on your crystal.

"The circuit diagram is shown in Fig 7.8 and is as simple as can be with just four major parts – a voltage regulator/switch, an RS232 interface chip, the 2313 and the R2R resistor network. The R2R network is connected to PORTB on the 2313, making it a simple D/A converter and makes it possible to output 256 voltage levels.

"Neither the resistor network or port drivers of the 2313 is of perfect linearity, but it works pretty well anyway. But you'll probably need a buffer stage as the output impedance is rather high (tens of kilohms in my case). The MAX603 handles the voltage regulation as well as the power-up/shutdown function, and is controlled by the DTR signal on the serial interface. So when you shut down the control program on the PC, the Mini DDS will be shut down for battery saving. Fig 7.9 shows a picture of the Mini DDS, built on a PCB that fits in a TEKO box.

"The control software user interface is illustrated in Fig 7.10, and this, together with the 2313 source code software, is available from reference [2].

"The software is written in assembler, as it is very short and it needs the speed in the main loop. This is the heart of the synthesiser. The rest of the code is the wave tables and the communication code. The phase accumulator uses 24 bits, which determines the resolution of the output frequency. Maximum available frequency and resolution are also dependent on your crystal frequency.

146

"The main processing loop of the DDS takes up just six instruction memory words and executes in nine clock cycles. During this time the processor:

(a) adds the 24-bit offset to a 24-bit accumulator;
(b) uses the top byte of the accumulator to look up the required output amplitude from an eight-bit look-up table with data for sine, triangle, sawtooth and square waves;
(c) outputs the result of the sine table to the R2R ladder output.

At a clock speed of 11.06MHz this loop executes in just 814ns.

Fig 7.9. Mini DDS PCB

"Communications with the DDS is achieved by serial communications using interrupts and based around the built-in UART. One of the compromises of the design is that the DDS loop cannot run when the AVR is processing the serial commands. However, this is a small price to pay for such a powerful and low-cost synthesised signal source.

"The analogue output from the DDS is generated by using an R2R ladder connected to the AVR's eight PORT B outputs. This creates an elementary high-speed DAC. Accuracy (and therefore, to some extent, spurious DDS outputs) depend on the accuracy of the R2R resistor values. Note that some filtering and buffering will almost certainly be required on the output of the R2R ladder but is not shown on the circuit diagram."

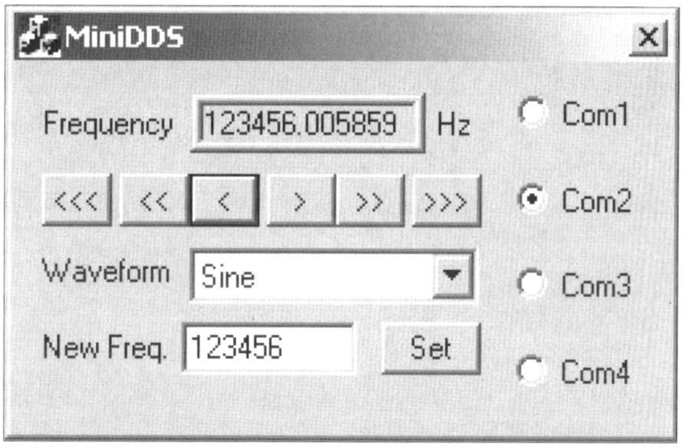

Fig 7.10. Mini DDS driver software

A Windows program is available from Jesper which drives the DDS from the COM port of a PC.

Resolution $= F_{CPU} = 150994944$ and $F_{OUT} =$ Accumulator $\times$ Resolution

Here, with a 11.059200MHz crystal, the resolution is 0.073242188Hz. To get an output frequency of 1kHz, we need to use a phase accumulator value of 0x003556 (13654 decimal). This gives an output frequency of 1000.048835Hz – good enough for most hobby work!

The communication is pretty simple and just allows you to set and read the phase accumulator value, as well as selecting the type of output waveform. It uses a 32-bit value for the phase accumulator to be compatible with the larger DDS circuit using the AD9832 chip.

## References

[1] 'The PicATUne automatic antenna tuner', Peter Rhodes, G3XJP, *RadCom* September 2000 to January 2001.

[2] Mini DDS: Jesper Conran website at http://www.myplace.nu/avr/minidds/index.htm.

# CHAPTER 8

# Digital signal processing

## Basics of DSP

Digital signal processing (DSP) is the technique of taking a signal, usually an analogue waveform, and performing mathematical operations such as addition and multiplication on a series of digitised samples. Many of the tasks that are performed by DSP would traditionally have been done in the past by analogue means with operations that might typically involve frequency conversion, filtering and demodulation – all of which occur in a conventional communications receiver. Modems for data modes are these days nearly always implemented in DSP, although analogue designs may have been used in the past for modes such as radio teletype (RTTY). The ST5 RTTY terminal unit was a classic example in its day of a modem constructed using analogue techniques.

A number of other specialised techniques have become available in DSP that were hitherto impossible to visualise. The classic example here is the fast Fourier transform (FFT) which will be covered later in this chapter. Its analogue equivalent can be roughly approximated by visualising a huge bank of hundreds, or thousands, of very narrow band-pass filters side by side in the frequency domain, starting at DC and going up to the highest frequency present, with an amplitude detector then sitting on the output of each filter. In the 'sixties the UK Foreign Office invented a data mode called *Piccolo* which transmitted one of 32 audio frequencies, each corresponding to a letter of the alphabet or symbol. The tones were decoded by making use of narrow high-$Q$ filters made from resonating reeds and this, along with similar copy-cat multilevel FSK systems, is probably one of the few analogue implementations of what is now a very simple task in DSP.

## Sampling

Obviously the first requirement is to turn the continuously varying input signal into a series of digital values that can be processed. We met A/D converters in an earlier chapter and some of these are perfectly suitable for DSP use; generally we need as high a frequency response as possible, coupled with the maximum resolution or dynamic range that can be achieved. What is possible with state-of-the-art (or perhaps more important to the

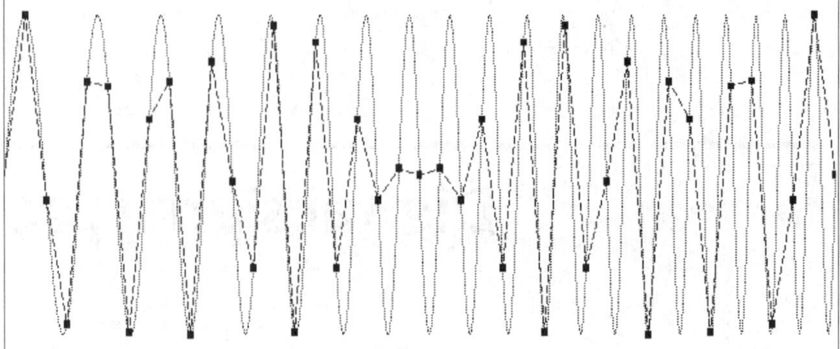

Fig 8.1. Sampling a swept sine-wave signal

radio amateur, low-cost) devices necessarily involves a trade-off in one or both of these areas, but technology and commercial interests are rapidly pushing forward the performance of A/D converters so that devices are available now for a few pounds that would have cost many thousands of pounds only a few years ago.

For sampling audio, and even low radio frequencies such as up to 25kHz, there is a wide range of custom converters in existence designed specifically for DSP uses. They include both A/D and D/A converters with selectable sampling parameters as described below. These devices are usually called *CODECs*, and are designed to interface directly to a microprocessor bus or custom serial interface.

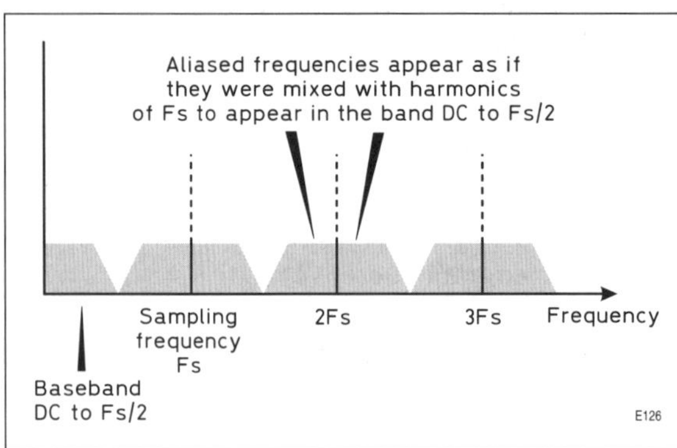



Fig 8.2. Spectrum folding due to aliasing

## Nyquist

The first hurdle to overcome when sampling a signal is the *Nyquist criterion*. Nyquist, a mathematician, proved theoretically that it is possible to perfectly sample a signal provided that the sampling rate (number of samples per second) is at least twice the highest signal frequency present in the input waveform. Fig 8.1 illustrates the sampling of a swept sinusoidal signal and shows what happens as the frequency of the sampling waveform goes above half the sampling rate. The sampled waveform, shown by the dots, appears to reduce again in frequency. This critical value of half the sampling rate is the *Nyquist frequency*, above which any input signal will be folded back into the fundamental band. This process is called *aliasing*, and in fact a whole band of frequencies either side of each of the harmonics of the sampling waveform will all be mapped to the wanted frequency range as illustrated in Fig 8.2. So it is clear that *all* components above half the sampling rate have to be cut out by an anti-aliasing filter before digitisation in the A/D converter. The anti-alias filter is usually of an analogue type, although switched capacitor designs are often used. If this latter solution is chosen, care is needed to ensure high-frequency switching signals do not leak into

**150**

the A/D converter input and themselves cause alias products which would appear as unwanted carriers or tones.

## Harmonic, or band-pass, sampling

Note that the Nyquist limit refers to the bandwidth, or information content, of the signal and not its absolute frequency. Looking at Fig 8.2 we see that it is quite permissible to sample a band-pass filtered waveform that encompasses one of the alias bands rather than the fundamental – for example one occupying a band 6kHz either side of an IF of 455kHz – at a rate that is dependent only on the bandwidth. If this is done then the A/D converter has to be physically capable of operating at the high frequency so that it can handle the waveform through its sample-and-hold circuitry. The overall effect is to mix down the actual input frequency by behaving as a 'pseudo' local oscillator at the nearest harmonic of the sample rate that will give a mixer product below the Nyquist limit. So now the actual sampling frequency has to be chosen carefully to ensure the same harmonic is used for the entire bandwidth and the converted spectrum does not 'fold back' on itself.

The 455kHz example given above, as it has a bandwidth of 12kHz, could theoretically be sampled at 24kHz. (We will have to assume that the anti-alias band-pass filter before the A/D conversion is perfect, and only signals from 449 to 461kHz can possibly be present. Needless to say, in practice this is an unrealistic requirement!) If the sampling rate is actually taken to be 24kHz then the nearest harmonic to 455kHz of this is the nineteenth which has a value of 456kHz – right in the middle of the pass band! So a new sampling frequency has to be chosen, remembering that this still must be higher than 24kHz. A suitable value might be 1/16 of the lower frequency limit of the input band, or $449/16 = 28062.5$Hz. A more convenient value close to this can then be tested – such as 28kHz. Now, the lower frequency limit is effectively mixed down to $449 - 16 \times 28$kHz $= 1$kHz, and the upper limit mixes to 13kHz, making use of the same 'harmonic'. Both these limits are 1kHz inside the 0–14kHz allowed, and this can be made use of in designing the cut-off band for the anti-alias filter. It now only has to produce its full attenuation outside the range 448–462kHz while letting 449–461kHz through.

By now it should have become obvious that the anti-alias filter is a very important part of any DSP system. CODECs all include internal anti-alias filtering by customised switched capacitor or similar filters. Usually, frequencies up to 0.45 times Nyquist can be dealt with, and the attenuation above 0.5 $Fs$ is below the resolution of the converter. Amplitude ripple in the pass band is also very low, and is usually specified in the CODEC's data sheet. Filter performance can sometimes be adjusted by user set-up instructions.

## Dynamic range

The next issue is that of *dynamic range*, or *sampling resolution*. All A/D converters are defined by the number of bits of quantisation they use, and

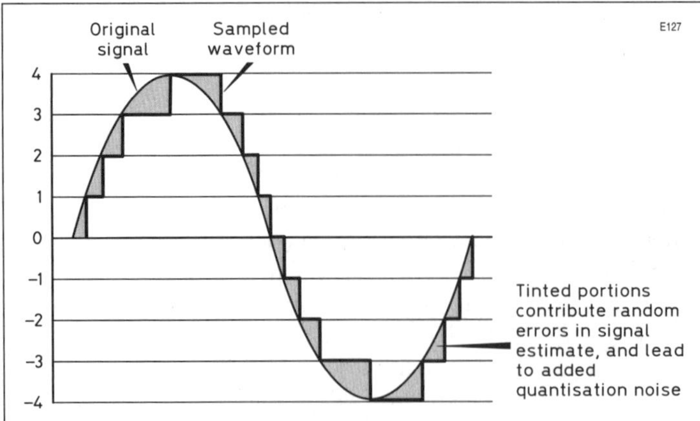


**Fig 8.3. Generation of quantisation noise**

consequently a sampled input waveform can only be approximated to a number of discrete levels. For an eight-bit converter 256 levels are available, while for a 16-bit converter this becomes 65,536 levels. Fig 8.3 illustrates the portions of the quantised waveform contributing to the error, and it can be seen that each individual quantisation point introduces a random amplitude error component. The result of adding all these successive quantisation error samples is to add a term that looks and behaves in a way identical to noise – it is referred to as *quantisation noise*. The level of quantisation noise is linearly related to the number of quantisation levels, leading to a simple logarithmic relationship between the number of bits and the amplitude. By expressing the amplitude of the quantisation noise in decibels relative to the RMS value of a *fully sampled sine wave*, it is possible to show that the signal-to-quantisation-noise ratio is given simply by:

$$S/N_Q = 6.N - 1.75 \text{ dB}$$

This equation only applies for a sinusoidal signal that exactly fills all the quantisation levels, and for typical signals such as voice and modem tones, the figure for $S/N_Q$ needs to be modified to take into account the peak-to-mean (or more correctly peak-to-RMS) ratio of the waveform of interest. Uncompressed speech normally has a peak-to-RMS ratio of around 10dB so for eight-bit quantisation the absolute maximum signal-to-noise ratio that could be achieved after A/D conversion followed by restoration is therefore:

$$6 \times 8 \text{ (bits)} - 1.75 - 10\text{dB (peak to mean )} = 36\text{dB (approx)}$$

The telephone service adopted eight-bit quantisation as a practically realisable number of bits for the network and, as this figure of S/N was not considered to be good enough for a toll-quality service, measures had to be taken to improve on this value. The solution adopted was to effectively compress the speech waveform to reduce the peak-to-mean ratio, raising the S/N to around 40dB. Rather than actually adding speech compression hardware, a non-linear quantisation was employed which has the same overall effect and allows hardware simplification. Two very similar techniques are used worldwide; in Europe and much of the world the quantisation equation is referred to as *A-law quantisation*, in the USA there are slight differences and the system is then called *μ-law*. Many CODECs for telecommunications offer both these quantisation types, as well as linear quantisation to different numbers of levels which can be chosen in their initial set-up.

For radio purposes a signal may already have a considerable amount of noise present – particularly when its fundamental signal-to-noise ratio is less that that due to the quantisation, so eight-bit conversion can often be employed for RF purposes if the highest S/N performance can be sacrificed. As a rule of thumb it is generally accepted that, for extracting signals out of noise, the inherent system noise due to RF front-end or antenna contributions has to occupy the first three to four quantisation levels of whatever degree of quantisation is finally adopted. So higher quantisation resolution does mean that signals can be dug further out of the noise, even when they are much worse than that due to the quantisation itself.

One final point that should be noted about quantisation noise is that it occupies the entire spectrum from zero hertz to the Nyquist frequency. So, if any digitised signal is subsequently filtered to have a narrower bandwidth, with a corresponding reduction in sampling rate (a process called *decimation* that will be covered later) then the quantisation noise is also reduced by this filtering – to below that of the original quantisation noise. In fact, it is possible to sample a signal at eight bits at quite a high sampling rate, then filter and decimate while expanding to 16-bit quantisation at a lower sampling rate purely by mathematical operations. This trick is one part of the fundamental design of a software radio; progressively reducing the bandwidth and sampling rate to get from a wide RF bandwidth down to a single audio channel.

The signal samples occurring at regular intervals are referred to as *time domain data*, as they represent the waveform over time – in the way an oscilloscope shows how a waveform varies over time. The other way of representing signals is in the frequency domain, where the horizontal axis represents frequency. The corollary to this view of the signal is that shown on a spectrum analyser display. We will meet the frequency domain later in the section on the FFT process.

## Frequency conversion and the numerically controlled oscillator

So-called 'frequency mixing' is actually a multiplication process rather than real mixing (which is simply addition of signals). New frequencies are generated when two sinusoidal signals are multiplied, and this falls out of the trigonometric identities:

$$2 \times \sin (A) \times \sin (B) = \cos (A - B) - \cos (A + B)$$
$$2 \times \cos (A) \times \cos (B) = \cos (A + B) + \cos (A - B)$$

or

$$2 \times \cos (A) \times \sin (B) = \sin (A + B) - \sin (A - B)$$
$$2 \times \sin (A) \times \cos (B) = \sin (A + B) + \sin (A - B)$$

By replacing $A$ and $B$ with terms for signal and local oscillator frequencies, the source of the sum and difference mixer products that always occur together becomes obvious. Note the relative signs of the two mixer products;

the first is positive, the second is minus depending on whether we use cos.cos or sin.sin. This little fact comes into its own later as a method of cancelling one sideband, but for a straightforward multiplication of one carrier by another in a mixer, any form of the above equations is applicable as phase has no meaning when not referred to any particular reference, so the cos or sin form is irrelevant and both mixer products will always appear simultaneously.

All conventional analogue mixers fail to meet this ideal multiplication in some way or other, and in most cases a large LO signal has to be used, meaning that all its harmonics and all the mixer products of these appear in the output waveform and have to be removed by filtering. Possibly the only analogue mixer that does act as a real multiplier is the transistor tree (Gilbert cell multiplier) as used in the ubiquitous MC1496 chip and many later-generation RF devices such as the NE612.

In DSP, multiplication can be perfect – it is only a mathematical operation after all! So now frequency conversion of a digitised and sampled signal becomes that of multiplying each sample value with that of a sampled sine wave at the local oscillator frequency. This LO is usually generated in a numerically controlled oscillator (NCO) which works as follows.

Values of sin and cos are generated for the required local oscillator frequency at the sampling rate of the input signal and *exactly* synchronised to it. The resulting pair of values at each step point is used as the input to the multiplication. Actual sin and cos generation is ideal but expensive in terms of computation time. So what is usually done is to maintain a table in memory with the values for sin and cos for a complete cycle at, say, $1°$ intervals. In this case there would therefore be 360 pairs of values stored but the quantisation of the angle will introduce some phase jitter in the look-up. Instead of 360 entries, if we make the table a binary power of two in length, (say 256 entries at intervals of $360/256 = 1.40625°$), we can see that the rollover from 359 back to $360 = 0°$ will be equivalent to the binary address pointing into the table overflowing from the 'all ones' state back to zero. Many DSP chips have 256-long sine tables built into them

This is best illustrated by example. In DSP it is standard practice to normalise all frequencies to the sampling rate, so that this assumes a normalised frequency of one. The input band up to the Nyquist limit will then always occupy values between zero to 0.5, and the LO will likewise fall within this range of numbers. This gives a clue as to how the look-up table is used. Assume a local oscillator frequency of 0.1 times the sample rate. For the very first sample, the first term in the look-up table is taken, which for a sine table is zero. Then for the next sample, the address pointing into the table is incremented by the normalised LO frequency, so for a table that is, for example, 256 samples long the next address would be at $0.1 \times 256 =$ number 26, the next address 51 and so on. Every time the address reaches 256 it wraps round modulo 256 to zero again, simulating the next LO cycle. The value of table lengths that are powers of two can now be seen – modulo 256 just means truncating all bits of the address higher that the eighth.

Note that we have now introduced a frequency quantisation term due to the rounding of the address, and also the rules of Nyquist apply equally to generation of the NCO waveform – no frequencies higher than half the sampling rate are allowed. In fact, aliasing is more obvious in frequency generation than it is for sampling: any frequency above half the sampling rate would necessitate a phase rotation per sample of greater than $180°$. Any angle greater than $180°$ is equivalent to a negative rotation of less than $180°$ so the net result is a falling frequency as the requested value rises.

The Nyquist limit is easily designed into (or out of) the system but the table length and rounding issues can be a problem. The effect of the frequency quantisation error is to introduce sidebands onto the pseudo-carrier that then appear as added noise – behaving more like phase noise this time. Suitable choice of table length can alleviate this, subject to available memory limits, but there is another trick of mathematics that can be applied to interpolate the table address to give a much higher resolution – this will be covered later where specific DSP routines are described.

## I/Q conversion

Frequency conversion in DSP can be made much more precise than in analogue mixers, and completely removes the need for filtering or removal of the unwanted 'mixer' product. The technique is to modify the NCO routine to generate two outputs per sample point, representing output waveforms $90°$ out of phase with each other. In other words cos and sin terms which we will label cos $(A)$ and sin $(A)$ here.

Two look-up tables are required for this process, or alternatively a double look-up with addresses corresponding to the $90°$ difference between sin and cos. The two sets of values are each multiplied by the input samples to give:

$$\cos (A) \times \sin (B) \quad \text{and} \quad \sin (A) \times \sin (B)$$

where $A$ is the numerically controlled oscillator and $B$ corresponds to the sampled input signal.

So far, all we have done for each sampling period is to generate the quadrature local oscillator by getting two entries from a look-up table and performing two multiplications, but we have now ended up with two values from the mixing process, corresponding to $90°$ phase-shifted terms. These are referred to as *I/Q samples*, which is short for *in-phase and quadrature samples*.

Now we come to an even more useful aspect of DSP. These I/Q terms taken together completely represent a frequency-shifted version of the input signal, and the actual frequency shift is irrelevant – it can be anything, subject to valid Nyquist values for the frequencies of course. It is perfectly reasonable and valid to frequency mix the input waveform with a local oscillator that sits in the middle of its pass band.

For example a sampled voice bandwidth signal of 300–3300Hz can be mixed with an LO of 1800Hz to give a 'folded' spectrum that goes from $1800 - 300 = 1500$Hz at one extreme, to zero frequency corresponding to 1800Hz in the middle of the audio spectrum, up to $3300 - 1800 = 1500$Hz

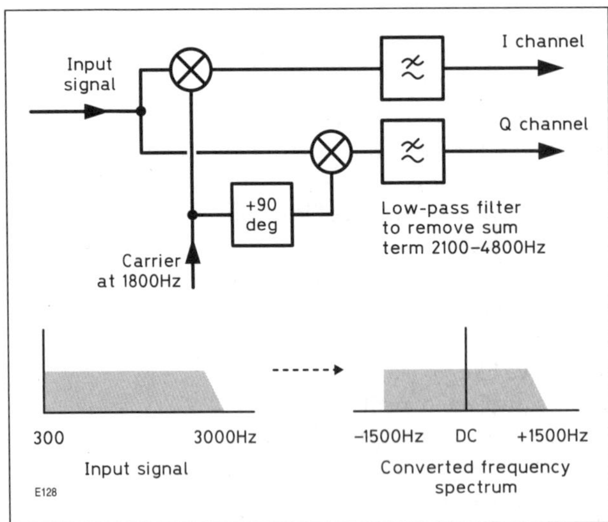



**Fig 8.4. Folding of the frequency spectrum in I/Q mixing**

at the other extreme. On its own this is pointless, as now we have mixed up the spectral components so they overlap each other. But, by virtue of having the I/Q pairs of samples, the spectrum folding can be eventually sorted out and unwrapped.

Furthermore, the I/Q samples now only have half the bandwidth of the input signal and range in frequency components from DC to exactly half the input bandwidth.

Fig 8.4 shows this I/Q mixing process graphically. An analogue equivalent to this can be seen in the 'third method' of SSB generation where a 0/90° phase-shifted voice signal is mixed with a physical 1800Hz local oscillator and low-pass filtered to 1500Hz, before being subsequently converted up to the wanted carrier frequency. Accurate matching of all paths was always needed for the third method, so this means of generating an SSB signal never achieved popularity. In the world of DSP nearly all processing is done with I/Q signals since these completely represent the entire input waveform. Each of the two values are independent and can take on both positive and negative values.

They can be represented by the vector diagram shown in Fig 8.5. where the I term represents the amplitude along the X or horizontal axis, and the Q term the amplitude along the Y or vertical axis. Mathematically this is referred to as *complex notation*, where the I component can be likened to the 'real' component of a complex number, and the Q component to the 'imaginary' term. There is little need to go into complex notation for describing most DSP procedures and that will be the approach taken here, but many DSP textbooks can make quite simple techniques seem a lot more complicated than they really are by adopting purely mathematical terminology and explanations.

This diagram in Fig 8.5 is often referred to as a *vectorgram* or *phasor diagram*. Apart from being a mathematical tool, the vector diagram showing I/Q signals can be generated quite easily as a real visual tool by taking the I and Q values, and using them to display a spot at the respective X and Y positions on a display screen.

The same result is obtained by converting the I/Q values to analogue in a pair of D/A converters and displaying the resulting analogue values on a the X-Y display of an oscilloscope after passing through a low-pass filter to get rid of the alias terms.

A signal at exactly zero frequency (derived from a sine-wave input with a frequency exactly equal to the local oscillator NCO) appears as a dot on this diagram. If the frequency is changed slightly this dot continuously rotates around the origin, or centre of the chart, with the direction of rotation depending on whether the frequency is higher or lower than the reference or centre frequency. The I/Q sampling mode of the PIC-based DSP interface described in Chapter 4 allows a vectorscope to be made. This operates at a fixed centre frequency of 1000Hz and the pattern is formed by plotting dots at the X and Y co-ordinates given by the I/Q values on a graphics screen.

To complete the description of I/Q representation, we need to know how to convert it back to understandable form. The amplitude of the overall waveform is generated by using the formula of Pythagoras to calculate the amplitude of a pair of vectors at right-angles (the hypotenuse of a right-angled triangle whose other two sides are the I/Q signals):

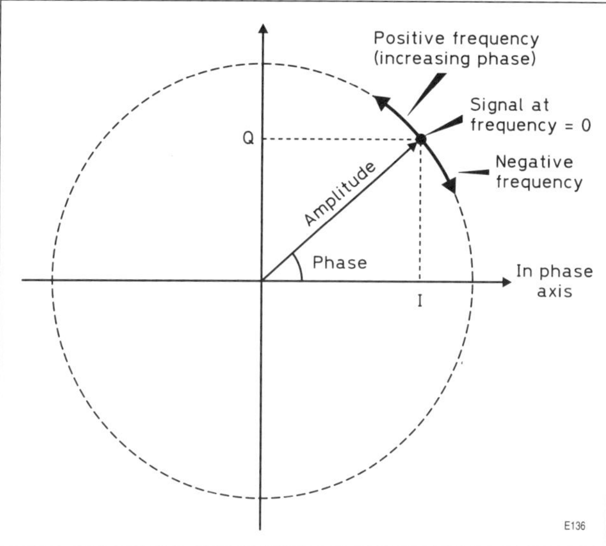


Fig 8.5. Vectorscope representation

$$A = \sqrt{(I^2 + Q^2)}$$

If the values of A are fed to a D/A converter and a loudspeaker we have an AM demodulator, with the real audio output derived directly from the down-converted I/Q samples.

The relative phase can be obtained from:

$$\theta = \arctan (Q/I)$$

but note that the full four-quadrant form of arctan needs to be employed to generate a phase value from 0 to 360°. In several high-level languages this appears as the ATAN2 function. In DSP chips there are other mathematical tricks for obtaining a sufficiently accurate approximation to the arctan function without the time-consuming effort of calculating it exactly. In some cases it is not always necessary to calculate the phase in order to make use of it – one example of this is FM demodulation.

## FM demodulation

Frequency is defined as the rate of change of phase, so if we continuously compare the instantaneous ATAN result with the previous one we can calculate the rate of change of phase, and hence the frequency. We now have a means by which we can demodulate FM from an audio tone (defined by the NCO centre frequency) without actually hearing any of that tone itself. Such a technique is ideal for slow-scan TV, for example, where we can recover a bandwidth of 1350Hz from an FM carrier centred on 1650Hz, in an SSB bandwidth of 300–3000Hz, even if the carrier tone swings below 1350Hz.

The difficulties with the ATAN function can be resolved for FM demodulation: instead of actually generating the arctan function then calculating its rate of change by comparing samples, the function itself can be differentiated mathematically – which gives an exact equation – then the result of the mathematical manipulation is incorporated in the calculation of the rate of change of phase. The mathematical derivation is beyond the scope of this book but results in:

$$d\phi/dt = \text{Instantaneous frequency}$$

$$= I_{(n)}.Q_{(n-1)} + Q_{(n)}.I_{(n-1)} / [I^2_{(n)} + Q^2_{(n)}]$$

where $I_{(n)}$ and $Q_{(n)}$ are the current I and Q samples, and $I_{(n-1)}$ and $Q_{(n-1)}$ are the previous samples. With four multiplications and a division, it is clear that this is a much easier calculation to perform in real time at high speed, and it is usually built into the software of many DSP-based radio receivers.

## Correlation

Correlation is a term that covers a very wide area of DSP and pervades the whole field of signal processing. It basically involves multiplying a signal by a reference waveform and then integrating, or averaging, over a number of terms. The result is often to extract the wanted signal from noise or interference.

Here we will perform only one of the simplest examples of correlation, that of mixing a signal with a sine wave to extract sinusoidal CW-type signals from noise. This example will be used to illustrate how DSP software is written in a normal programming language. The software can run in any high-level language on a PC, even with most 16-bit languages in a DOS environment, and is straightforward enough that results can clearly seen even if the answers are just printed to the screen! The process involves a multiplication of the digitised input signal by sin and cos terms of a numerically controlled oscillator, then averaging the results over many terms. An A/D converter that can feed digitised audio samples to a PC is first needed, and the PIC-based serial A/D converter design of Chapter 4 is ideal for this job. This can take a signal received in an SSB bandwidth and send eight-bit samples at 10kHz sampling rate to the PC over the serial port operating at 115200 baud – well within the capability of most PCs within the last decade. As an alternative, for programmers who can write for the Windows operating system, the sound card can be used as the means to get audio samples – an introduction to Windows programming and accessing the sound card appears in Chapter 10.

For the purposes of this section, the software will be written in *pseudocode*. This is a fictional programming language where commands and instructions are kept simple and equivalent to those available in all high-level languages – so conversion into your language of choice is kept straightforward. Apart from setting up arrays, all other bits of set-up and initialisation – such as those for serial ports and the display – have been left out as this part of the code has no effect on the operation of the routines shown and is

**Table 8.1. 'Pseudocode' listing of a programme to correlate a received signal with a locally generated numerically controlled oscillator then integrate over successive samples**

```
Dim InputData(1024, 1) ;Set up an array to store 1024 successive I/Q converted A/D samples

SampleRate = 10000          ;For the Serial A/D, use 11025 for soundcard etc.
CentreFrequency = 900    ;Signal centred on 900 Hz tone
NormFreq = CentreFrequency / SampleRate
WaitFor Interrupt
;................
SerialInterrupt        ;Every new A/D sample that appears causes a jump to this routine
  N = GetValue
  InputData(HeadPointer , 0) = N *  COS(2 * Pi * Phase);Write I data into circular buffer
  InputData(HeadPointer , 1) = N *  SIN(2 * Pi * Phase);Write Q data into circular buffer
  HeadPointer = (HeadPointer + 1) MOD 1024 ;Update circular buffer pointer
  Phase = FRACT(Phase + NormFreq)                ;Ready for next NCO sample

  If HeadPointer = 0 then          ;Only need to do this every 1024 samples
     SumI = 0 : SumQ = 0              ;Clear out any residue from last time
     for X = 0 to 1023       ;Use 1024 successive samples, can be any lower number
         TailPointer = (HeadPointer + 1 + X) MOD 1024 ;EARLIST sample in memory,
; is that immediately AFTER the current  head pointer.   Move forwards to the latest.
         SumI = SumI +  InputData(TailPointer, 0)
         SumQ = SumQ +InputData(TailPointer, 1)

     next X           ;1024 summations of I and Q now complete
     SumI = SumI / 1024   :   SumQ = SumQ / 1024       ;Normalised values
     SignalPhase = ATN2(SumI, SumQ)    ;Full four quadrant phase calculation
     Print Amplitude  ,   SignalPhase     ;We can do better than this, but its enough!
  end if
FinishInterrupt
```

very dependent on the target language. If the pseudocode listings that follow appear to show any relationship to Power Basic, that is purely intentional!

The listing (Table 8.1) illustrates a number of processes common to the majority of DSP software. First, a few variables are set up for the sampling frequency and the local oscillator (equal to the tone frequency that will be converted to zero frequency). The normalised frequency term is calculated to be used in the NCO generation later.

The whole programme is written such that one calculation loop is performed every time a new A/D sample arrives – by definition this must be at exactly the sampling rate. Either the software itself will generate the A/D conversion or, for example, characters arriving on the serial port at a constant rate will generate the processing event. The programme has to make sure that all input samples are correctly taken, at the right time, and that none get lost. In a dedicated DSP unit this could be generated by a hardware interrupt linked to the A/D converter. In a large operating system like Windows, the samples are just 'handed' to the software in blocks supplied from the sound card, these are sequentially worked through and the software then waits for the next block with no attempt to do this in real time. Correct timing of the input samples is handled only within the sound card.

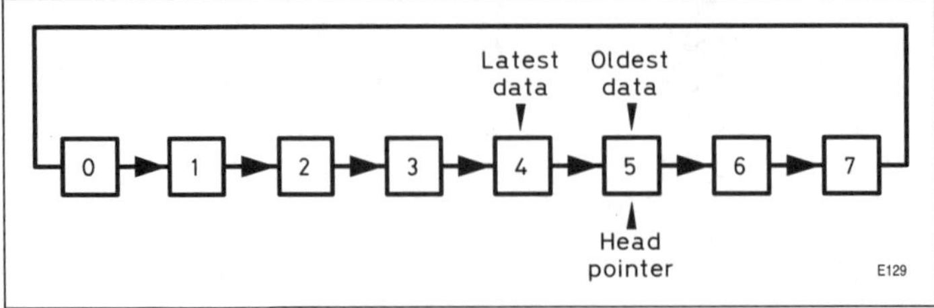



**Fig 8.6. Circular buffer addressing. Each new sample is stored at the head pointer location and overwrites the oldest sample. The head pointer is then incremented modulo 8**

The first job is generation of the local oscillator by an NCO routine. For simplicity, the NCO is generated by calling the sin and cos functions, and in fact, when writing in a high-level language running on a PC, this is usually the fastest method as the PC's maths co-processor contains very fast and efficient sin and cos generation routines. Within a custom DSP chip it is more usual to employ look-up tables rather than trying to calculate the sin and cos values. For each sampling period a variable (called `Phase` here) is continuously incremented by a fixed amount equivalent to the normalised local oscillator frequency. As it corresponds to an angle, the value has to increase progressively then jump back to zero.

There are many ways to generate this angle – adding a fixed number of degrees, then taking the result modulo 360 to give an angle between 0 and 360 is probably the most familiar way, but most software does not work with degrees. Calculating the angle directly in radians would be the ideal solution, but the mod ($2\pi$) function is not mathematically possible as $\pi$ is not a whole number, so we instead just take the fractional part of the number, which has the same effect as if we took the result modulo 1. The value obtained is then converted to radians for the sin and cos calculation by multiplying the resultant value between 0 and 1 by $2\pi$. This solution to generating the NCO also means that the value by which the phase has to be incremented for each sampling period to give a correctly sampled LO waveform is given exactly by its normalised frequency. There is also the advantage that many DSP chips work with binary values normalised to represent numbers between $\pm1$. So, when using these devices, the calculation almost becomes as simple as adding the phase increment, then taking the result 'raw' and ignoring any binary overflow, exactly as would be the case for integer binary arithmetic.

Every A/D sample is multiplied by the current instantaneous values for the I and Q terms of the local oscillator and the pairs of values of the frequency-converted waveform are stored in a circular buffer so that the last 1024 values are always available. A circular buffer is used so that the latest set of samples are always available in memory, with each new sample overwriting the oldest one. Fig 8.6 shows the circular buffer addressing graphically. For this routine it would be just as easy to read a block into a fixed array starting at the top each time, and the use of circular buffers may appear a little long-winded when programming in a high-level language. However, array pointers and addressing will always be required whatever storage routine is adopted, as well as a means of moving the data around, and a circular buffer is one of the fastest and most memory-efficient means to achieve this. Custom DSP chips include a set of commands and addressing modes that make access to a circular buffer automatic and virtually transparent to the calling routine.

Next, we test for the point at which a complete set of 1024 points of new data are in the memory – this is simply done here by testing for the input memory pointer to equal a value of zero. This event will occur every 1024 cycles of the sampling frequency, and if this is at a rate of 10000 samples per second, the test will give a valid result every 10.24ms, or at approximately 9.77Hz. For other sampling rates the frequency would change accordingly. When this test is valid, we sum all 1024 I values and all 1024 Q values, and divide by 1024, giving the average value over the 10.24ms period. The final stage is to calculate the amplitude and phase of the result and display them.

So what does this result actually represent? Consider an input signal that is a sine wave of arbitrary phase and frequency exactly equal to the NCO. If it were exactly in phase with the NCO I channel, the resulting `SumI` value would be a maximum and `SumQ` zero, similarly if the input waveform shifted by 90° the `SumQ` would be at a maximum and `SumI` zero. So we have generated the amplitude and phase of the input signal – which could have been done on each individual sample – but the averaging process has done a lot more than this. If the frequency varies very slightly, the result seen on a per-sample basis would hardly vary. However, over 1024 successive samples each multiplication would produce an error from the result expected if the frequency were correct. The effect of these errors is to reduce the sum value quite rapidly as the frequency alters. In fact, at a tuning error of exactly half the repeat period, or 4.88Hz, the resulting `SumI` and `SumQ` will be zero. So, as this situation can occur at tuning errors above and below centre frequency, we have generated what amounts to a 9.77Hz band-pass filter followed by a power and phase meter. We have a 9.77Hz bandwidth AM receiver.

As mentioned earlier, this process of multiplying an input waveform by a locally generated signal, then averaging (integrating) over a period is called *correlation*. As a tone detector it is not perfect – valid results will also occur (although the values will be lower) for tuning errors that sit at all odd harmonics of the bandwidth, but as an illustration of how DSP works it suffices. In fact this routine makes a good tone detector that can operate in parallel with other more complex DSP processes acting on, for example, voice signals that can be used to extract very low-level signalling tones concealed behind the main signal.

Correlation forms a major part of many modem and data signalling processes. The locally generated signal is not limited to just a sine wave; it can be any waveform at all, to match the shape of quite complicated transmitted signals. One example used by amateur radio operators for propagation monitoring is correlating an SSB receiver's output with a locally generated signal that is a pre-stored representation of a signal that ramps from 200 to 2500Hz in 20ms – a 'chirp' signal. Then, the correlator output is at a maximum each time one of the many chirp sounder signals that sweep at a rate of 100kHz/s and are transmitted from around the world passes through the receiver pass band. By modifying the above programme so the summation is done for *every* received sample rather than just each time a new block

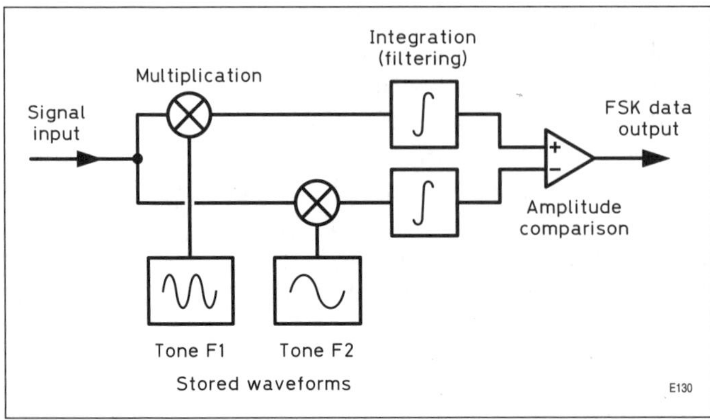


**Fig 8.7. FSK demodulator using a pair of correlators**

appears, it is possible to determine the exact timing of the chirp to an accuracy equal to one sample period. This is made use of in the Chirpview software written by G0TJZ and G3PLX [1].

It is also possible to correlate against several locally stored signals each time, rather than just the one. Then, by comparing each of the corresponding summed values, the maximum is obtained from the one corresponding to the stored waveform that most matches the transmitted waveform. By storing two tone frequencies, for example, an FSK demodulator is the result as shown in Fig 8.7.

## The fast Fourier transform

If we want to look at a sampled input signal and determine its frequency content – for spectrum analysis, for example – one way to do this would be to generate a multiple correlator routine as described above with a set, or bank, of sinusoidal reference waveforms of different frequencies stretching from the lowest one equal to the frame repetition rate up to the highest possible limited by Nyquist. A convenient grid would be to have frequencies that are integer multiples of the block rate. For the example above, all the tones that are multiples of 9.8Hz up to a maximum of 5kHz would be tested by the correlator, 512 separate correlations in all. That way, the relative levels (and phases) of each of the sum terms directly gives the spectral content of the input waveform to a resolution of 9.8Hz. This process was first invented by the mathematician Joseph Fourier as a mathematical transform on continuous data and used across a huge spectrum of engineering and scientific fields, not just for frequency/spectral analysis. In fact the Fourier transform was first used in calculating heat flow and also appears in calculations with phased array antennas.

However, the Fourier transform, or the *discrete Fourier transform* (DFT) as it should be called when used with fixed sampled data, is a very computationally intensive routine. For each of the 'tones' there are $N$ complex (I/Q pairs) of multiplications and additions, so the relatively complexity is proportional to $N^2$. Multiplication is easily the most time-consuming mathematical function in the process and dominates the calculation time. For the 1024-long block used above, over 1 million complex multiplications would have to be performed every 10.24ms – something that is possible with the latest PCs, but certainly not with the computers that were in existence at the time when Fourier analysis started to be widely used, around the 'sixties.

However, at that time two mathematicians, Cooley and Tukey, noticed that there was a lot of redundancy in the DFT process and that the process could be considerably simplified by applying certain limitations. They worked out that, provided the block length was an integral power of two, such as 8, 16 . . . 256 etc, the DFT process could be progressively broken down into a series of 1-bit transforms, and the results re-assembled. The net result is then to eliminate all the redundancy in the DFT process and means that the resulting fast Fourier transform (FFT) now has a relative complexity of $N.\log_2 N$ complex multiplications rather than $N^2$ so the calculation proceeds considerably faster. As an example, the 1024-point DFT would have needed 1048576 complex multiplications, whereas by using the FFT this is reduced to 10240 – speeding up the process 100 times. The actual mechanics of the FFT process are rather complicated but for anyone interested is explained in nearly all DSP texts – some in less mathematical terms than others. Complete descriptions are given in references [2] and [3].

The listing shown (Table 8.2) is one FFT routine written in Power Basic (or pseudocode). This particular routine was taken originally from a Fortran listing written almost at the time of the FFT's invention, and only required a few modifications to the syntax to convert to the Basic language. It has the property of storing the output (frequency domain) data in the same array as the input (time domain) data. So if the original data is still needed after FFT processing, a copy will have to be kept in another array. As the data array is complex (with real and imaginary terms), the input samples are stored in the real (or in-phase) half of the array, with the equivalent imaginary (or Q) locations being set to zero. After the FFT routine has completed, the first half of the array up to an index value of $N/2$ contains the complex coefficients of the spectral frequency-domain components. The second half of the array contains the same data in a mirror image – this corresponds to frequencies that would (or could) have been above the Nyquist frequency between 0.5 to 1.0 times the sampling rate. Each array location containing the I/Q amplitude coefficients for a frequency term is usually referred to as a *frequency bin* – as in dustbin.

## Negative frequency

If, instead of filling the input data array with raw A/D samples and stuffing the Q channel with zeros, we fill the array with the I and Q components of a digitally down-converted I/Q signal, we get a very interesting and useful result – the output array no longer contains duplicate frequency spectra in mirror image. Now, the lower half of the array contains one frequency spectrum from DC to half the sampling rate as before, but the other half of the array, from index $N/2$ to $N$, contains the lower sideband from the frequency conversion, also from DC going to $Fs/2$ in the other direction – negative frequencies. We have managed to extract and separate both overlapping spectra from the conversion. Now, since we have full non-overlapping coverage from $-Fs/2$ to $+Fs/2$ we have full spectral information for a band equal to the sampling rate. This concept appears to fly in the face of Nyquist, but is a natural artefact of being able to make use of negative frequency. Many professional engineers, let alone amateurs, feel uncomfortable with

**Table 8.2. Fast Fourier transform routine converted to PowerBasic from Fortran**

```
sub FFT(dat(1) , nn%) ;Enter with data stored in array called dat,  and size of data block
'data format  (re,im,re,im,....) starting at 1
    shared pi              ;3.14159265358979323846264383279 (approximately !)
    n% = 2 * nn%
    j% = 1
    for i% = 1 to n% step 2
        if j% > i% then
            swap dat(i%) , dat(j%)
            swap dat(i% + 1) , dat(j% + 1)
        end if
        m% = n% \ 2
        while m% >= 2 and j% > m%
            j% = j% - m%
            m% = m% \ 2
        wend
        incr j% , m%
    next i%
    mmax% = 2

    while n% > mmax%
        istep% = 2 * mmax%
        theta = 2 * pi / mmax%
        wpr = cos(theta)
        wpi = sin(theta)
        wr = 1
        wi = 0
        for m% = 1 to mmax% step 2
            for i% = m% to n% step istep%
                j% = i% + mmax%
                tempr = wr * dat(j%) - wi * dat(j% + 1)
                tempi = wr * dat(j% + 1) + wi * dat(j%)
                dat(j%) = dat(i%) - tempr
                dat(j% + 1) = dat(i% + 1) - tempi
                dat(i%) = dat(i%) + tempr
                dat(i% + 1) = dat(i% + 1) + tempi
            next i%
            wtemp = wr
            wr = wr * wpr - wi * wpi
            wi = wi * wpr + wtemp * wpi
        next m%
        mmax% = istep%
    wend
end sub
```

the concept of negative frequencies but it is a real mathematical concept, and of practical use in DSP where I/Q processing can really keep the two halves of the spectrum separate ready for reconstruction later.

### Windowing

One difficulty with using the FFT to calculate the spectrum of discrete sampled data is what happens at the sampling boundaries. If we take an arbitrary number of samples, say 1024, of a random input waveform then the

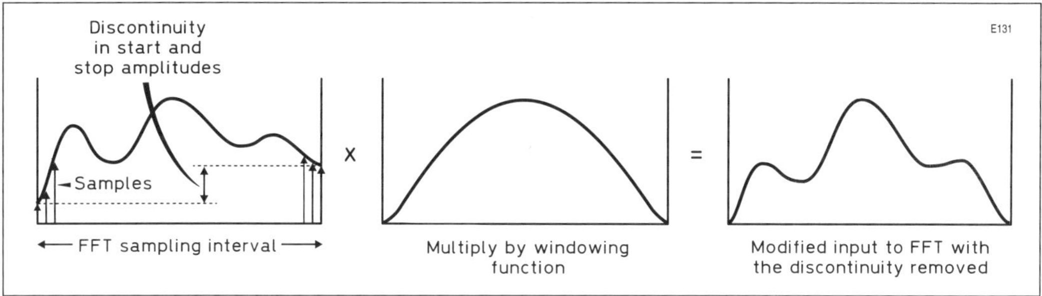


beginning and end of the sampled waveform will be discontinuous. This is illustrated in Fig 8.8. The effect of this is to cause spectral leakage across frequency bins, where the step transition causes a random raising of the level of many bins. It can be shown mathematically that at worst, approximately 40dB signal/noise ratio is all that can be achieved with straight eight-bit sampling.

The spectral leakage raises the noise floor to 40dB below the maximum level a single carrier at maximum amplitude could give. The way round this is to multiply the input data by a set of window coefficients so that the first and last samples in the block are reduced practically to zero – the ones in the middle of the sampled block are virtually untouched and those either side scaled proportionately. If this is done, the spectral leakage across all bins is reduced but at the expense of effectively widening individual bins. The bandwidth of each bin is no longer that of the block sample rate but is now wider. Spectral leakage is moved into adjacent bins rather than in to all of them in a random fashion.

The amount of leakage that can be reduced, and the resulting new effective bin width, is a function of the window type. FFT windowing is a huge area, the subject of many past studies, but for our purposes we only need to consider two or three different window types, each with different trade-offs between signal/noise ratio and effective bin width.

Two of these are given in the software listings shown in Table 8.3. It is also possible to apply a window to the calculated frequency domain data after the FFT process, and in certain cases this can be a more computationally efficient way to proceed.

Another way of looking at the need for windowing is to remember that the FFT calculates the spectrum of a block of signal in isolation, not a continuous signal. Imagine instead of a block of data, a burst of single tone equivalent to the block length. If this is fed into a spectrum analyser the abrupt transitions at beginning and end lead to sidebands either side – similar to key clicks. If the burst included several carriers, each would have its own key clicks which would overlap and make it difficult to reconstruct the original signal.

The solution is to round off the edges of the burst by reducing their amplitude – analogous to a key-click filter. The effect on the FFT is to eliminate the unwanted products, but since we have to round off the burst, its length becomes shorter, and a shorter burst means less resolution in the

Fig 8.8. Discontinuity in sampling, and windowing

**Table 8.3. Two routines to window the time domain data before the FFT process in order to reduce spectral leakage**

```
sub BlackHarrisWind (dat() , n%)     ;enter with data array (re,im, re, im …..etc)
  shared pi
  for k% = 0 to n% - 1
  locdat% = k% * 2 + 1
    theta = 2 * pi * k% / n%
    dat(locdat%) = dat(locdat%) * (.42323 - .49755 * cos(theta) + .07922 * cos(2 * theta))
  next k%
end sub

sub HammingWind (dat() , n%)
  shared pi
  for k% = 0 to n% - 1
    locdat% = k% * 2 + 1
    theta = 2 * pi * k%/ n%
    dat(locdat%) = dat(locdat%) * (.54 - .46 * cos(theta))
  next k%
end sub
```

frequency domain. By starting with a longer block in the first place, and overlapping blocks, we can get back to the resolution we wanted in the first place albeit with increased processing complexity.

### The inverse FFT

To complete the discussion of the FFT process, we note that the same routine can be used for conversion in the opposite direction, from frequency domain back to time domain – this is achieved just by changing the sign of one of the multiplications in order to get the output results in the right order. Otherwise the FFT process is completely reversible, converting from time to frequency and back again.

The ability to swap from frequency domain to time and back suggests one way of processing an audio signal. If we perform an FFT on an audio signal in real time, then modify the amplitudes of the resulting frequency bins in line with the wanted filter response, then do an inverse FFT (an IFFT) on the result, we generate a time-domain signal that has been filtered. The filter 'shape' is that of the weighting applied to the frequency bins after the first FFT process, Fig 8.9. The resulting time-domain values from the IFFT can be sent to a D/A converter to recover a real electrical, and possibly an audio, waveform. By adjusting the coefficients that are applied to the frequency bins we can tune or adapt the filter at will. This method of producing a digital filter is a very complicated and rather inefficient way of achieving an audio filter, and there are many better ways of digital filtering a signal. But it does allow arbitrary and unusual frequency responses to be obtained; modern computers and DSP chips are more than adequate in supplying the processing power needed.

Some amateur software for the PC using a technique like this allows an arbitrary frequency response to be drawn – using the mouse – which is immediately converted to a filter response and the processed audio is replayed though the sound card speakers [4].

# Digital filtering

One method of digitally filtering a signal has been described. However, there is a much more generic, computationally efficient and faster technique that can be used for the majority of filtering functions.

A basic digital filter, and the simplest to visualise, involves taking a number of successive samples of an input waveform, then multiplying each one of these by a coefficient and summing the results. For the next input sample, all the previous samples are moved back one, the earliest is discarded to make room for the latest, and all are multiplied by the coefficients and summed again. The process is repeated for each sample and is shown in Fig 8.10. This is known as a *finite impulse response* (FIR) filter as the output is generated uniquely from a finite length of the input data. Generating the coefficients for the multiplication is more complicated but does only have to be done once if the filter is not to be changed or tuned during operation. Usually fixed coefficients are stored in programme memory or calculated from first principles whenever the DSP chip is first initialised. So how do we generate the coefficients to give a particular filter shape, for a low-pass or band-pass response for example?

A straightforward method is to make use of the fact that time-domain and frequency-domain data are interchangeable and can be converted from one to the other via the Fourier transform. An ideal low-pass filter in the frequency domain is a step response – a band-pass filter is often described as a 'brick wall'. If we take the Fourier transform (or strictly the inverse FT as we are going from frequency to time) of the wanted filter shape we end up with a series of time-domain samples, which by a happy coincidence just happen to be the set of coefficients needed for the multiply-and-add digital filtering. The same problems with spectral leakage apply so we usually need to include a window function as well. For the 'perfect' step or brick wall responses there is no need to perform an actual DFT on the shape, as the result is a well-known mathematical relationship (the sin $(X)/X$ shape and the coefficients can be calculated from this equation at initialisation, or on-the-fly during operation if the filter response is to be altered.

The filtering cannot be perfect, and the sharpness of the frequency transition depends on the length of the block of data chosen. This, in turn, has

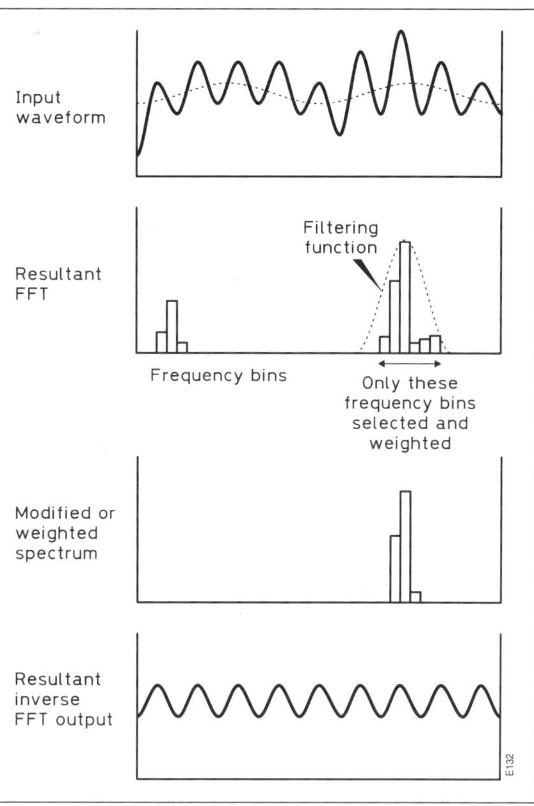


Fig 8.9. The FFT/IFFT digital filtering technique

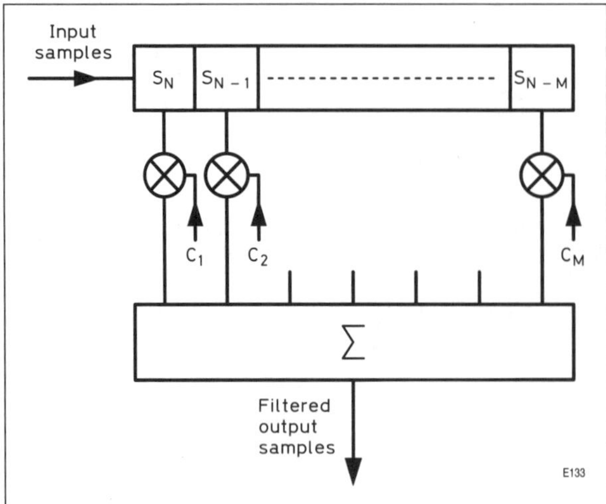


**Fig 8.10. Digital filtering, the FIR filter**

to be chosen to allow reasonable processing speed and memory requirements to be met. Obviously, the faster DSP processors with more memory will allow longer filter lengths and hence more precise filtering to be achieved. Normally block lengths of 32 to 1024 are chosen and meet the requirements of most audio processing needed by amateurs. However, the filter shape can be much more flexible than anything produced by analogue means, the only limitation being governed by the sampling rate/block length criteria. Another disadvantage to long filter lengths is the time taken for a signal to pass through them. If the output is needed quickly for a fast turn-around, two-way system such as AMTOR, a fairly 'soggy' filter response may have to be tolerated since we cannot afford the time to let the filter fill up.

One incidental advantage to the FIR filter is that it is inherently linear phase if the tap array is symmetrical about the centre – which is usually the way they are designed. As each output sample is a linear combination of a fixed number of previous samples, the time shift introduced by the filter must be fixed (equal to the block length/sampling rate) so the phase response cannot be anything other than linear. Linear phase is always a good thing to aim for on filters designed for audio as it minimises ringing and sounds better to the ear. The finite response nature means that classic filter shapes such as Butterworth, Chebyshev etc cannot be implemented perfectly, although good approximations to their shapes can be obtained by Fourier analysis.

## Circular buffers in digital filtering

Now the value of using a circular buffer that we first met in the correlation example should become more obvious. Every new input data sample requires accessing all the last $N$ samples to do the filter function. So, by starting at the buffer location immediately in front of the point where the latest sample is stored in the buffer, which contains the oldest sample, then working forwards, all $N$ samples can be accessed quickly and in the correct sequence without having to actually move any data in memory; only the pointers are updated on each pass though.

It is clear now that as we need to be able to do $N$ complex multiplications and a summation for every input sample, where $N$ is the number of historical samples at any time in the block – the filter length. This is quite a lot of processing. All custom DSP chips include a command like MAC – multiply and accumulate – which performs one coefficient multiplication and sums with the previous value. This whole multiply-and-add command usually takes up one DSP processor clock cycle, so the whole filtering process takes $N$ clock cycles per input sample. Pointers need to be updated, but in most DSP chips this can occur in parallel with the MAC command. So we are beginning can get an idea of the processing speed required for DSP. For filtering a 10kHz sampled signal, a 256-long filter will need a processor that can do $256 \times 10000$ multiplications per second, or 2.56MHz assuming a hardware multiplier that takes one clock cycle to complete. This is hardly fast by modern standards and the DSP chips of a decade or more ago run happily at 60–80MHz. Modern technology allows hundreds of megahertz so we can see DSP on audio signals is going to be a relatively straightforward task.

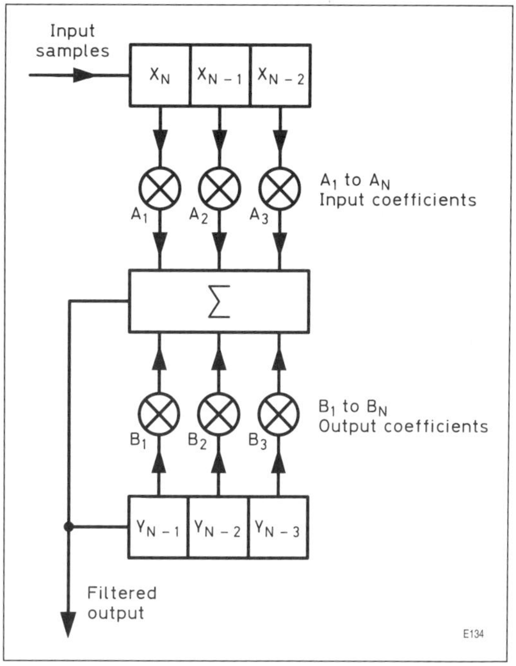

### The IIR filter

Instead of just multiplying and summing input samples, another type of digital filter reads the output of the MAC command and sums these with another set of multiply-and-accumulate commands derived from the last few samples of the output data, which is stored in another buffer. This is illustrated in Fig 8.11. Now, the data within the filter routine depends on *all* the previous samples, and the new result can affect, theoretically at least, all future output samples – hence it is known as the *infinite impulse response* (IIR) filter and reflects the situation in analogue filters. IIR filters are characterised by needing very much shorter block lengths for a given sharpness than FIR designs, so they can generally be used to much higher frequencies, but are more difficult to design and get going. They can easily become unstable if the coefficients are not exactly right, and one particular problem can occur due to summation of the mathematical rounding errors present in integer maths, the MAC command can be very badly affected by biased rounding errors. This is to a large extent alleviated with modern DSP chips by using floating point arithmetic – but these still do not automatically ensure perfect IIR filter results, and so most amateur designs keep to FIR filters for simplicity and because they do work as they should.

Input samples

$X_N$  $X_{N-1}$  $X_{N-2}$

$A_1$  $A_2$  $A_3$  —  $A_1$ to $A_N$ Input coefficients

$\Sigma$

$B_1$  $B_2$  $B_3$  —  $B_1$ to $B_N$ Output coefficients

$Y_{N-1}$  $Y_{N-2}$  $Y_{N-3}$

Filtered output

E134

**Fig 8.11. Infinite impulse response (IIR) digital filter**

# Decimation

An audio input signal, say from an SSB receiver, will usually be sampled at a rate not much higher than Nyquist allows. Typically, SSB audio is sampled at 8kHz and in some cases as low as 7.2kHz. However, if our aim is ultimately to filter this signal to a much narrower bandwidth, for signal analysis or narrow-band signalling for example, the results will come out of the digital filtering process still at the same sampling rate as they went in – this is a whole tenet of the DSP process, the sampling rate dictates everything. Now, a narrow-band signal, say only 30Hz wide (such as for the PSK31 data mode, for example), can be represented quite well at a sampling rate of twice this bandwidth – much lower than the existing value.

If the sampling rate can now be lowered, the processing time that is permissible for each sample can go up as there is now more time between each sample. Digital processing for data communications such as error detection and clock recovery can be much more complex than the frequency conversion and filtering just described, and the ability to do this at a reduced sampling rate considerably simplifies the task. The technique of reducing the sampling rate is called *decimation*.

One way is to down-convert to zero frequency using I/Q mixing techniques, then low-pass filter the result to half the wanted signal bandwidth. Remembering that the I/Q mixing technique folds the spectrum over itself, low-pass filtering this has the effect of applying a band-pass filter to the original signal that is centred on the LO frequency. Now, we know the upper frequencies that can be present are dictated by the digital filter response, and that the zero frequency signal is now being *over-sampled*, ie sampled at a rate very much higher than its Nyquist frequency. What we can now do is just to take one sample out of every few, ensuring that the resultant new sampling rate is sufficient to meet the Nyquist criteria for the filtered signal.

This can be shown by making use of a typical example. Assume a signal from an SSB receiver (300–3000Hz) is digitised at 8kHz sampling rate. We want to derive a 30Hz bandwidth signal centred on 1kHz to pass on to a demodulator or FFT routine. First, we down-convert to zero frequency by making use of an NCO at 1kHz, synchronised with the input samples. This generates an I/Q signal, still sampled at 8kHz. This is then passed through a FIR filter with a cut-off of 15Hz – half of the wanted signal bandwidth. From experience we know that this degree of filter shape will need a length of around 1024. We also know that a signal of 15Hz can be safely sampled at any frequency greater than twice this, or at any value higher than 30Hz. So if we take one sample out of every 128 input samples we can lower the sampling rate to $8000/128 = 62.5$Hz. (Although we could theoretically work with half this value, 31.25Hz, the leakage and imperfections in the digital filtering would lead to alias products. These are greatly reduced or even eliminated by adopting the higher rate.) We simply throw away the other 127 samples out of every 128 that come out of the digital filter.

Now for a little nicety that speeds up the whole DSP process no end! If we are going to throw away 127 samples out of every 128, then why calculate them in the first place? Digital filtering is quite time consuming and we don't want it to use up so much processor overhead that the other more complex demodulation routines suffer. The output of the multiply-and-accumulate command at the end of a complete pass through the 1024 buffer samples contains the filtered output sample. However, since we want this only once out of each 128, why not just wait until then? Provided all the input samples are correctly stored in the circular buffer at the primary sampling rate, the software just has to test for every 128th character and perform its MAC calculation on the buffer contents at that single event. The rest of the time can be spent doing other tasks.

## Ultra-narrow-band processing

There is no reason why the decimation process cannot be continued further – the same technique can be applied again to further reduce sampling rate. From the example given above, exactly the same code and filter coefficients can be re-used to give a new sampling rate of $62.5/128 = 0.488$Hz and a signal bandwidth of 0.234Hz. (The filter coefficients that were calculated for a 15Hz low-pass cut-off at 8000Hz sampling rate now give a new cut-off frequency, proportional to the change in sampling rate. Here it is 128 times lower at 0.1172 Hz). If a 1024-point FFT is now performed on this data, the frequency spectrum of a narrow band centred on 1kHz (or whatever the NCO is tuned to) can be resolved to a resolution approaching 0.000477Hz. In practice, due to requirements for FFT windows the resolution bandwidth will be more like 0.001Hz but that is still very, very, narrow. Such bandwidths require extreme frequency accuracy to be of any use, and clock-rate error in the A/D converter can show up here, even assuming frequency errors in the receiver frequency conversion are eliminated.

This mix-filter-decimate procedure is the backbone of much DSP-based narrow-band signal processing software, especially that using dedicated DSP chips and low-cost solutions. However, DSP software for spectral analysis on the PC does not always use this technique. PCs are characterised by having huge amounts of memory and fast maths co-processor chips, which, coupled with the fact that audio sampling rates are relatively slow compared with processor clock means that decimation is sometimes unnecessary. In their signal analysis software, a number of software authors instead perform a huge FFT on the raw 8kHz sampled data. A 262144-point transform $(2^{18})$ or 1M point $(2^{20})$ is quite feasible with the later generation of PCs, allowing direct viewing of signals in bandwidths of 0.03Hz or less. Reference [5] gives details of a narrow-band spectral analysis package.

# References

[1] Chirpview is available from G0TJZ's website: http://www.asenior48.freeserve.co.uk/chirpview.html.

[2] *The Scientist and Engineer's Guide to Digital Signal Processing*, Steven W Smith, California Technical Publishing. Individual chapters of the book can be downloaded free of charge from www.dspguide.com. A very easy-to-read introduction to DSP.

[3] *A Course in Digital Signal Processing*, Boaz Porat, John Wiley & Sons Inc. A more theoretical and mathematical treatment of DSP.

[4] SR5 Spectrum analyser available from www.ar5.com.

[5] I2PHD's website at http://www.weaksignals.com/ for Spectran, Argo and other DSP software for the sound card.

# CHAPTER 9

# Practical DSP hardware

In order to make use of the various techniques available to us with digital signal processing, we need to look at the hardware and platforms available to run the software and then present or make use of the data. For the radio amateur, there are two routes that can be used, and each has its advantages.

## Platforms for practical digital signal processing

Where the main function of DSP is for analysis and decoding of signals, such as for data modes and spectrogram-type software, there is a lot to be said for using the processing power available in the modern PC. Furthermore, high-resolution graphics are an integral part of all modern home computers and the Windows operating system means that the full graphics potential can be realised for real-time display of signal spectra or other parameters. The sound cards present in all modern machines make digitisation of audio straightforward and the huge processing power of machines running with a 32-bit or 64-bit bus at clock speeds of hundreds of megahertz is hard to beat. This is the subject of the next chapter.

However, where the requirement is for a small stand-alone unit, such as an audio filter or noise-reduction unit for connecting to the output from a receiver, dedicating a PC to the job is not usually practical. This is not to say it isn't done. The SR5 software described in reference [5] of Chapter 8 is an example of a user-defined audio filter with audio input and output via a PC sound card, although more generally a small dedicated unit is wanted that can be just switched on, and is then up and running immediately. Enter the dedicated DSP chip.

## DSP chip

Before the days of PCs running at hundreds of megahertz, a custom digital signal processor chip was the only practical way to perform any real-time processing at audio frequencies. Within this chapter we will concentrate primarily on one brand of DSP chip, the 56000 family by Motorola. This is not to say that it is any better than those from other manufacturers, but it is arguably one of the easier to understand and to begin programming from first principles. A low-cost evaluation module, the 56002EVM, based around this chip, made huge inroads into the use of DSP by radio amateurs in the 'nineties. That particular module is regrettably now obsolete, and second-hand versions are snapped up even before they appear on the market. There is an updated version now available with a later DSP chip which can be used with most of the original software, although software modifications to suit the upgraded functionality and hardware will be required.

A DSP chip is similar to any other microprocessor in the way it addresses memory, using several working registers and an arithmetic and logic unit for manipulating the data and moving it around. However, the difference comes in the types of instructions that are available, as well as some extra hardware included within the arithmetic unit. All the usual microprocessor-type instructions such as fetch from and store to memory, and programme flow control such as jumps and branches, are present and programmed in the same way, so all the 'normal' jobs required of a processor can be performed. DSP chips always have a larger set of internal registers than microprocessors – these are separate from the memory map and are addressed directly by the machine-code instructions, so maximising speed by not having to address separate memory at all. Also present in a DSP chip is a hardware parallel multiplier since, as we saw in the last chapter, many DSP operations include repeated multiplications. The multiplier can usually perform a complete multiplication on two words taking the full capacity of the processor bus, and give a double-precision answer in just a couple of clock cycles. The other main areas of difference in the hardware is the memory addressing. Classic microprocessors (and the PC) make use of the same memory and address bus for both data and machine code – known as the *von Neuman architecture*. For this every data fetch requires an absolute minimum of two clock cycles, one to read and decode the machine instruction, the next to get the data it has to act on. Many commands take multiple clock cycles to complete so this architecture is not the best for speed, but does have to advantage of circuit simplicity as only one bank of memory has to be addressed and decoded externally.

DSP chips employ a different architecture, where the memory for storing the programme code is kept separate from that for the data, with separated address and data busses for each. This is called the *Harvard architecture* and leads to a more complex hardware solution to manage separate busses, but now means that machine instructions and data can be accessed together. To further complicate matters, it is usually the case that the same physical address and data busses are used for accessing external memory, which is separated by the use of control lines. By ingenious use of multiple clock phases both sections of memory can be addressed within a limited number of clock cycles over the same bus. This is done to minimise the pin count on the chip package and simplify PCB layout. We don't need to concern ourselves with how the DSP processor manages this bus-sharing task, however, as it is sufficient to know that there are functionally two separate areas of memory operating independently of each other.

The PIC processor described in earlier chapters also makes use of the Harvard architecture. Employed to speed up the processor to allow one, or in some cases two, clock cycles per instruction, there is no need for careful bus management on the PIC. All memory is internal and completely separate busses can be used with all connections inside the chip package.

The DSP chip contains a certain amount of random access data memory on board which can be addressed even more quickly than external memory with its bus multiplexing, so allowing the most time-critical routines to be speeded up still further. There are also sections of read only memory containing data such as sine and cosine look-up tables, which can be switched in or out depending on the programmer's requirements.

One final area of note is the specialist addressing modes allowed. We saw the value of the circular buffer in the last chapter, and special registers with a number of specific machine-code commands and addressing options are available to set up and address these, while their use remains transparent to the routine that uses them by not having to separately calculate the address pointers.

In most DSP chips there are many internal pathways between the registers and arithmetic unit, and for many arithmetical commands, data can be moved around within memory simultaneously within the one instruction, provided they do not both need to address the same memory area or the same registers. For instance a circular buffer pointer can be updated whilst a multiplication is underway using the data stored in two registers. This allows programme memory to be conserved as all the commands are able to be stored together in one programme memory location.

# The Motorola 56000 DSP family

The first DSP chips were 16-bit integer devices, but 16 data bits do not allow very high precision in repeated mathematical operations such as multiply-and-accumulate operations, and their application was limited. However, modems using them were feasible, as well as some simple real-time audio processing tools such as DSP filters and add-on dynamic noise reduction units.

To remedy this, Motorola introduced a DSP chip making use of a 24-bit wide data bus with 24-bit registers, some of which could be doubled up, allowing 48 bits of resolution. Several devices in the 56000 family exist, with differences in the hardware included on each model to allow for different users requirements. However, all share a similar basic architecture as shown in Fig 9.1.

The basic processor core features:

- Operation at up to 20 million instructions per second (MIPS), making use of a 40MHz crystal
- Single-cycle $24 \times 23$-bit parallel multiply-accumulator
- Highly parallel instruction set with unique DSP addressing modes
- Zero-overhead nested Do loops
- Fast auto return interrupts

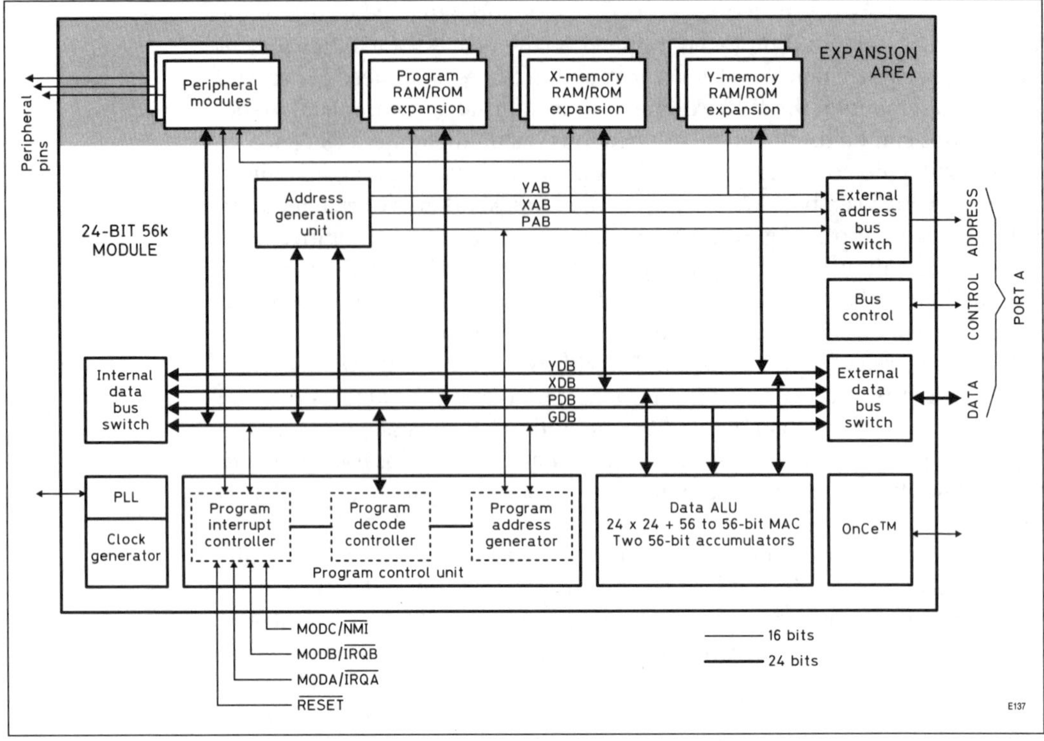


**Fig 9.1. General 56k DSP chip architecture**

- Fully static logic with operating frequency down to DC
- Very-low-power CMOS design
- STOP and WAIT low-power standby modes

The DSP56002 device features:

- $512 \times 24$-bit wide programme memory
- Two $256 \times 24$-bit data memory
- Two $256 \times 24$-bit data ROM (sine and cosine tables)
- Full-speed memory expansion port with 16-bit address and 24-bit data busses
- Synchronous serial interface port
- Serial communications interface (asynchronous) port
- 24 general purpose I/O pins
- 24-bit timer event counter
- On-chip emulator for unobtrusive, full-speed debugging
- Optional programme security features
- PLL-based clock with wide input frequency range multiplication and power-saving clock to reduce clock noise

### Address generation unit

One of the most important parts of the 56000 architecture is the wide variety of addressing modes available to simplify and speed up the common DSP tasks. Up to 64k of memory in the X bank as well as another parallel

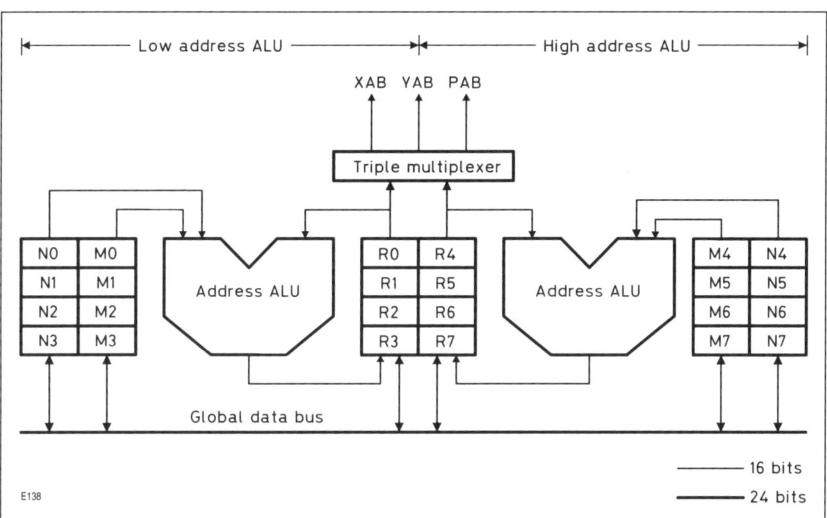



**Fig 9.2. 56000 address generation unit**

64k in the Y bank can be addressed by making use of 16-bit wide address registers. It is also possibly to directly access the programme, or P, memory. Fig 9.2 shows a diagram of the address generation unit (AGU) integral to the DSP56000 family. There are eight triplets of 16-bit wide registers and each triplet can be set up to perform linear, modulo and reverse-carry arithmetic, and operates in parallel with other chip resources to minimise address generation overhead. All three registers of each triplet are available individually for general-purpose 16-bit storage if they are not needed for addressing.

A few examples of typical instructions illustrate best how the AGU is used. In all cases the type of addressing, as specified by the command syntax, is automatically placed in the associated M register by the assembler/compiler:

1. MOVE A1,X: (R0)

   Uses the single R0 register from the first triplet to point to a location in X memory. The contents of accumulator A1 are moved (or copied) to this location in X memory.

2. MOVE B0,Y1: (R1)+

   The R1 registers point to a location in the Y memory bank. The contents of accumulator B0 are moved to here, and the address in R1 is then automatically incremented by one to point to the next memory location.

3. MOVE Y0, Y: (R3)-

   Moves the contents of register Y0 to the Y memory address stored in R3, then decrements R3 by one to point to the previous location.

4. MOVE X: -(R5), B1

   Decrements R5 by one *before* generating the address.

These first examples are all linear addressing modes, where the associated N registers are not used (and are available for general-purpose storage).

The following examples show some of the more complex modes possible that use the N register as well.

5. MOVE Y1,X: (R6+N6)
Moves Y1 to the address in X memory formed by adding the R6 and N6 register contents.

6. MOVE X1, X: (R2)+N2
As before for direct addressing with post increment, but now the address in X memory is post-incremented by the value in N2.

7. MOVE X: (R4)-N4, A0
Point to the memory location in X defined by the value in R4 and move its contents to accumulator A0. Then decrement R4 by the value stored in N4.

The last three examples of offset addressing are defined by storing the values 0xFFFF in the associated M register.

The next example shows modulo addressing as would be used with circular buffers. The value of the modulo (minus 1) is now stored in Mn, and similar offset addressing modes as illustrated above are used.

8. MOVE X0 , X:(R2)+N2 (with M2 containing the modulo value minus one)
The address in X memory is formed by adding the contents of N2 to R2 then taking the result modulo (M2+1). Thus the address generated wraps round and is constrained to lie between the values of N2 and N2 + M2.

## Arithmetic logic unit

The ALU contains hardware and interconnections for adding, multiplying and shifting data, as well as several general-purpose registers for feeding into these processes. Rounding logic is also incorporated to minimise errors caused by truncation of the results of calculations to fit into the 24-bit wide data storage. A block diagram of the data ALU is shown in Fig 9.3. Four 24-bit wide registers, X0, X1, Y0 and Y1 can form the input to the multiplier and the result can be stored or added to one of the two 56-bit accumulators. 56 bits are needed since, although a multiplication of two 24-bit numbers will give a 48-bit result, in the MAC command the results of many multiplications are added and this can easily exceed a 48-bit value. Hence 56 bits are allocated. Special commands are used to normalise a 56-bit result to generate a standard 24-bit result. By eliminating truncation of the intermediate multiplications within a MAC instruction, maximum accuracy of the final MAC summation is guaranteed so the DSP tasks that rely on this process will give the expected results.

Within fixed-width DSP numbers are not usually specified as integers, although the machine stores numbers as signed two's complement 24-bit values as a number between $-2^{23}$ to $+2^{23} - 1$ or approximately $\pm 2$ billion. Instead, a fractional representation of numbers is adopted where the values are scaled to fit within the range $-1$ to (just under) $+1$. The fractional representation is derived by dividing the 24-bit number by $2^{23}$. The use of fractional numbers makes the maths inside the DSP software and processor much easier to visualise (sin and cos can be represented by their proper values for example). The processor automatically places the decimal point correctly when 48-bit values, or higher, are generated, and subsequently converted to 24-bit values. For example, the un-normalised value in the 56-bit accumulators described above can actually take on a value between $-128$ to $+127.999\ldots$ This fractional representation is only applicable to the ALU part of the DSP where the mathematics of DSP tasks are actually performed. Conventional integer representation is still required for tasks such as memory addressing and control of peripherals, and instructions allow both types to be dealt with unambiguously. The programmer does have to take care of which data type is in use though, especially if converting one representation to another as may be required when outputting the results from a calculation (in fractional format) to byte-wide data for a communications interface, or into 8-bit to 16-bit integers for a D/A converter.

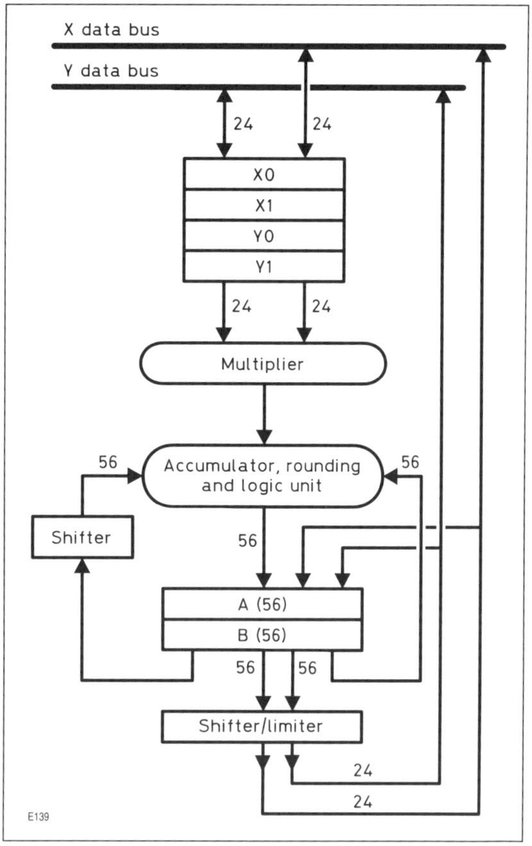

A worked example of a 56000 DSP routine serves best to show how the ALU operates: Table 9.1 shows the code for a FIR digital filter that filters I/Q data stored in X and Y memory respectively, according to filter coefficients stored in Y memory at location `filcofs`. A circular buffer has the length defined in `filsize` (all numerical constants are defined earlier in the set-up code where they are given named values). Whenever this routine is called once per sample the R1 register used as general-purpose storage contains the address of the latest data sample. R2 has been pre-loaded with the filter size. The line numbers in the listing are only for reference; they do not appear in any real assembler listing.

The first line loads the filter size (from R2) decremented by one into the N2 register ready for modulo addressing.

Line 2 clears the A accumulator ready for the multiply-and-accumulate to follow.

Line 3 shows how certain types of command can be executed simultaneously; as well as clearing accumulator B, the first filter coefficient in Y memory is loaded into R4 in a parallel move operation.

Line 4 uses the L syntax to refer to X and Y memory together, referred to as *long memory addressing*. This mode performs the dual function of moving X(R4) to register X0 and Y(R4) to register X1.

**Fig 9.3. 56000 data arithmetic logic unit**

| Table 9.1. DSP56000 code for a finite impulse response filter | | | |
|---|---|---|---|

```
1)    lua     (R2)-,N2            ;N2=filsize-1, loads N2 with updated offset
2)    clr     A
3)    clr     A,B   Y:filcofs,R4              ;zero A,B; also point to selected filter
4)    move    L:(R1)+,X                       ;get first I,Q pair into A+B together
5)    move    Y:(R4)+,Y0                      ;get first coeff
6)    do      N2,_end
7)      mac   Y0,X0,B   Y:(R1),X0
8)      mac   Y0,X1,A   X:(R1)+,X1   Y:(R4)+,Y0   ;do the filter
9)   _end
10)  macr    Y0,X0,B                         ;finish the filter
11)  macr    Y0,X1,A   Y:decmax,R4   ;
```

Line 5 gets the first filter coefficient from the filter table, using the pointer loaded in line 3. This illustrates one other area that programmers need to take care over. The data from a parallel move cannot be used immediately in the next instruction – instead there has to be some intermediate instruction code(s) to allow the internal pipeline to complete all its operations.

Line 6 sets up an operation that repeats all operations between the do command and the _end label, by the number of times stored in N2. Remember that N2 is the modulo count for the circular buffer which in turn is equal to the filter length minus one, so this loop will be executed exactly filsize – 1 times.

Line 7 is the multiply-and-accumulate commands for I data , acting on the current values stored in the X0 register multiplied by the filter coefficient in Y0 and storing the result in accumulator B. The parallel move in line 7 loads the next data value into X0, ready for the next iteration and based on the post-increment in line 4. The result is not available until three instructions later (when the loop returns to this same instruction).

Line 8 is the Q data multiplication by the same filter coefficient in Y0 as was used for the I data and storing the result in accumulator A. Two parallel moves here update the X1 register with the next Q sample while post-incrementing the pointer ready for the next iteration, and also getting the next filter coefficient to Y0.

Line 9 defines the end of the loop which is passed through until all but the last single pair of I/Q data and the associated filter coefficient have been multiplied and added.

Lines 10 and 11 perform the final MAC on the last I and Q samples respectively, but this special command also normalises the potentially 56-bit result in the accumulators back to a standard 24-bit value. The filtered result is available in A and B accumulators for whatever further processing is needed.

## Using the 56002EVM

Once we have written the DSP code for our target processor, we then need to get in into the DSP hardware, then debug and test the code. The description that follows is specifically for the now-obsolete 56002 evaluation module but the general principles are common to all DSP evaluation systems.

The 56002EVM is supplied with all the necessary software to get it up and running, as well as example code that can be adapted to give all the set-up and initialisation instructions. Such code is always going to be specific to the hardware in use. The most important area is often that for setting up the CODEC, setting the sampling rate and audio gain settings. One of the example programmes which is supplied with the 56002EVM, called 'PASSTHRU.ASM', is a complete listing including all set-up instructions that just takes an audio input from one, or both, of the line inputs, and passes it back out to the line outputs. While it has no DSP function in its own right, this little programme is absolutely invaluable for making the first tentative steps towards manipulating the data. It even includes a subroutine at the end of the listing called process_stereo which, as supplied, does absolutely nothing and just returns – it is in here that your first practice attempts at doing something with the audio samples can be added.

## Assembling the code

Once the .ASM file has been written, the first stage is to assemble it into machine code ready for downloading to the hardware and Motorola have supplied a straightforward DOS/command prompt programme to do this part of the task. The command line assembler called '56000ASM' is invoked by typing its name followed by the name of the source .ASM file. In practice, several command tags need to be included to control assembler output and a typical instruction to assemble the file 'PASSTHRU' is:

```
56000ASM -a -b -1 PASSTHRU
```

If all is correct (and if this file is exactly as supplied, it should be), a message will appear saying that assembly was successful. A number of new files will have now been generated, all with the same base name followed by different file extensions. The machine code is stored in the file with the .CLD extension.

If there are any errors in the source code, the .CLD file will not be generated. Instead, a description and the line number of the error will be written to an error file, eg PASSTHRU.ERR. The descriptions are quite concise, and usually point directly to the line of assembler that contains the error. Of course, this error checking cannot correct incorrect programme structure or functioning, but it can pick up most typing errors and, more significantly, all illegal operations and addressing modes.

## Downloading to the DSP hardware

The 56002EVM module is supplied with an RS232 serial link for downloading the operational code. A specific driver and debugging programme for the Windows operating environment called 'EVM56KW' is now used to drive the module via this interface. Other DSP cards may use the parallel interface which is generally somewhat faster than the serial one, or the USB interface, or the DSP card may be supplied as a plug-in card for the PC. Whatever form the DSP hardware takes, software will always be supplied with it to make the assembling and downloading tasks as straightforward as possible.

Once EVM56KW is invoked, the name of the .CLD file is entered into the box and the download function started. A progress bar shows the downloading process, and when this is complete the DSP card is running. It's as simple as that! The software allows multiple .CLD files to be loaded, such as filter tables that are generated separately if these are not specified as include files in the original source code. But there is more to EVM56K than just a downloader as it also contains many debugging tools for testing and help in getting your code operational. If the DSP software is not doing what it is supposed to, the code can be single stepped, line by line, and the contents of every register and sections of memory are visible at each step. Registers and memory can be changed at will, and programme flow can be diverted, for example to get out of 'stuck' loops and illegal functions.

## Using Flash ROM

Using EVM56K is a rather long-winded method of downloading code to the hardware – it can sometimes take several minutes. For proven code that you just need to get up and running immediately, there are two options available. Both methods require the addition of a 29C256 Flash ROM chip – containing 8k bytes of non-volatile memory – into the 28-pin ROM of the socket on the 56002EVM PCB. For a dedicated task that will run as soon as the module is powered up, or reset, the machine code can be directly stored in this Flash memory in a defined structure. A boot programme within the 56002 chip itself automatically downloads the contents of this ROM to the correct location in the DSP memory and starts running the code as soon as the download is complete. This only takes a fraction of a second and the results appears to be that the DSP system is running as soon as it is turned on. This immediate download is the preferred method for single-task dedicated jobs such as software receivers, audio filters and the like.

Loading the contents into the Flash memory in the correct format is made straightforward by a public domain utility called 'FLASH' which is available from reference [1]. The .CLD file is loaded into the EVM56K software as normal, but not run. Then file FLASH.CLD is loaded onto the end of the DSP code and the end address in bytes is inserted into register R1 using the debug tools. When this combination is run, the DSP code is blown into the Flash ROM ready for downloading next time the DSP card is reset. Full instructions are provided with the software.

The other option is a sort of half-way house between instant access to a single programme and the flexibility of using EVM56K. A utility written by Peter Martinez, G3PLX (although broadly compatible versions written by others also exist) works as follows. A loader programme called 'PLXBOOT.CLD' is saved in Flash memory as detailed above – this runs as soon as the 56002EVM is booted up and monitors the serial port input for incoming data in the correct format. On the attached PC, a programme

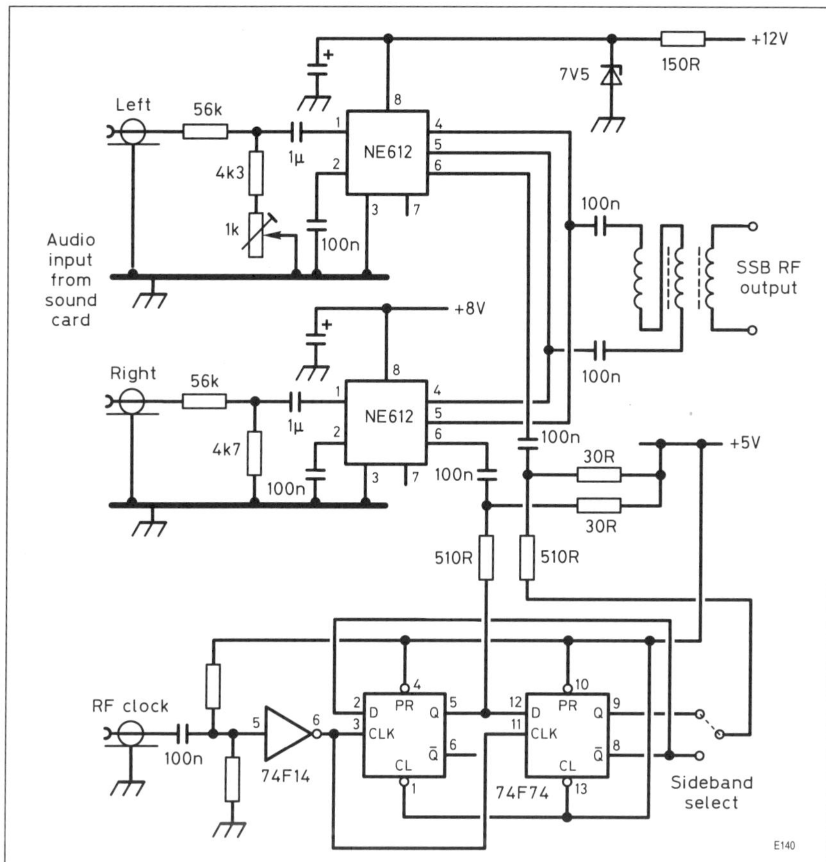



**Fig 9.4. Circuit diagram for I/Q up-converter**

called 'EVMLOAD' is run, followed by the .CLD file name, eg EVMLOAD PASSTHRU, at which point the contents of the .CLD file are uploaded to the DSP card via the serial link with error checking. The process is considerably quicker than going via EVM56K, and furthermore can be automatically invoked by building the source code or instructions for EVMLOAD into any user software, or by calling the loader directly from the programme. All code and instructions are available from reference [1].

## Phasing SSB generator

The listing in Table 9.2 was written by Peter Martinez, G3PLX, for directly generating an SSB signal at RF. The DSP code serves the same function as the audio 90° network, usually constructed from op-amps and close-tolerance components such as the polyphase network and others. It was written for the sound card but the routine can be adapted for any DSP card. The routine takes an audio input and generates I and Q audio channels, which are then connected to mixers driven by a quadrature local oscillator signal at the carrier frequency. The circuit of a suitable up-converter is shown in Fig 9.4. The result is a single sideband output RF at the LO frequency.

**Table 9.2. Listing of DSP code for generating the audio phase shift for a phasing-type SSB generator**

(First the Hilbert transformer array)

```
const filsize=79;
     hilbert:array[0..filsize-1] of extended=  {100-3900Hz, unwanted sideband  -70dB}
(
-4.42695301937857E-04, 1.67113161892970E-07,-5.84056554099099E-04, .69758824038231E-07,
-9.53526423361772E-04, 8.89408808871749E-07,-1.46227490318296E-03, 1.30587843953471E-06,
-2.14255405309416E-03, 1.64359475954973E-06,-3.03113766834784E-03, 1.82856752954826E-06,
-4.17043518526583E-03, 2.09966944102538E-06,-5.61016681715643E-03, 2.43741687373789E-06,
-7.41083969128908E-03, 2.86204580954991E-06,-9.64869830345094E-03, 3.02602387210887E-06,
-1.24247641710582E-02, 3.35546916410343E-06,-1.58812223256461E-02, 3.11595936939441E-06,
-2.02296744714943E-02, 3.33561855543618E-06,-2.58089096096512E-02, 3.00143211246981E-06,
-3.32026373168333E-02, 2.59458196245899E-06,-4.35174797055993E-02, 2.29710708936001E-06,
-5.91353542527600E-02, 1.85085624239585E-06,-8.62724013162506E-02, 1.28645138782781E-06,
-1.47770892289600E-01, 7.56638196995445E-07,-4.49393158790826E-01,-7.07106781186667E-01,
 4.49393158790826E-01,-7.56638196995445E-07, 1.47770892289600E-01,-1.28645138782781E-06,
 8.62724013162506E-02,-1.85085624239585E-06, 5.91353542527600E-02,-2.29710708936001E-06,
 4.35174797055993E-02,-2.59458196245899E-06, 3.32026373168333E-02,-3.00143211246981E-06,
 2.58089096096512E-02,-3.33561855543618E-06, 2.02296744714943E-02,-3.11595936939441E-06,
 1.58812223256461E-02,-3.35546916410343E-06, 1.24247641710582E-02,-3.02602387210887E-06,
 9.64869830345094E-03,-2.86204580954991E-06, 7.41083969128908E-03,-2.43741687373789E-06,
 5.61016681715643E-03,-2.09966944102538E-06, 4.17043518526583E-03,-1.82856752954826E-06,
 3.03113766834784E-03,-1.64359475954973E-06, 2.14255405309416E-03,-1.30587843953471E-06,
 1.46227490318296E-03,-8.89408808871749E-07, 9.53526423361772E-04,-4.69758824038231E-07,
 5.84056554099099E-04,-1.67113161892970E-07, 4.42695301937857E-04
);
```

Now a type declaration which is self-evident:

```
type Taudio=record
            left,right:smallint        {16-bit signed binary}
            end;
```

Next the SSB function, which takes a (stereo) audio input and returns a stereo result. The two sound card outputs go to I and Q DBMs, perhaps with a little bit of amplitude balancing to take out gain tolerance. The left stereo input ends up as the USB and the right stereo input as LSB, so it will even do ISB. A Pascal 'const' is a persistent local variable that keeps its value between calls to the function, but is not accessible outside the function. A 'var' is a transient local variable on the stack.

```
function SSB(input:Taudio):Taudio;
const  filptr:cardinal=0;     {i.e. an integer}
var I,Q,U,L:extended;      (floating-point numbers}
    tap:cardinal;
begin
  U:=input.left/32767;
  L:=input.right/32767;
  Ifil[filptr]:=(U+L)/2;
  Qfil[filptr]:=(U-L)/2;
  filptr:=(filptr+1);
  if filptr=filsize then filptr:=0;     {Ifil and Qfil are circular buffers}
  I:=0;
  Q:=0;
  for tap:=0 to filsize-1 do
```

**Table 9.2 *(continued)***

```
begin
  I:=I+hilbert[tap]*Ifil[filptr];          {+45 degree Hilbert filter}
  Q:=Q+hilbert[filsize-tap-1]*Qfil[filptr];  {-45 degree "backwards" Hilbert filter}
  filptr:=(filptr+1);
  if filptr=filsize then filptr:=0
  end;
  SSB.left:=round(I*32767);
  SSB.right:=round(Q*32767)
end;
```

The code is written in Pascal (with a few notes on variable types), and only shows the part of the function that processes each audio sample; all the other set-up instructions will be hardware and operating system dependent. If using a sound card, it should be set up for 16-bit stereo operation. As it stands the programme is intended for 8kHz sampling rate, but if this is changed, the frequency response just scales in proportion. The filter array is a modified Hilbert transformer, in which the phase shift is $45°$ over the audio band rather than $90°$. This is used to process audio for the I channel. The same array is used back-to-front to do a $-45°$ shift over the same band and that is used for the Q channel. Using a $\pm45°$ phase shift is hugely better than having a $90°$ Hilbert in one leg and a delay-line in the other. The filter is all-pass within the limits imposed by its length and the Nyquist criterion.

# References

[1] The Tucson Amateur Packet radio group, TAPR, have an extensive collection of 56002EVM software, as well as a collection of software for other DSP chips, on their website: www.tapr.org.

# CHAPTER 10

# Interfacing the sound card under Windows

## The event-driven model

One day it happened. It had to happen. Suddenly I found myself sitting in front of my PC, staring at a blank screen, with the keyboard waiting for me to type some arcane instructions to Windows, in order to persuade it to display a pretty panel and to accept orders based on the fact that some button had been pressed (or 'clicked', as the jargon dictates).

I had been up to that day a die-hard PC-DOS and microcontrollers programmer, whose motto was: "Real programmers don't use GUIs" where 'GUI', for the uninitiated, stands for 'graphical user interface', the Holy Grail sought first by the Menlo Park labs of Xerox, then by Apple and finally by some Redmond firm, producer of operating systems for the IBM (and compatible) PC. However, progress is progress, and it was time to say goodbye to my principles.

Enter the wonderful world of the event-driven model of programming. This is the first mental obstacle the would-be Windows programmer has to face. The so-called procedural model is the earliest that the apprentice programmer learns in his life, and the most intuitive. A programme is just a sequence of instructions, where the CPU is commanded to perform some tasks in the exact order the programmer had in his mind. Simple and beautiful. But there are some drawbacks.

What in a nutshell you do is to trade flexibility for simplicity in writing your programme. To see why, let's make an example taken from our hobby, ie amateur radio and electronics in general. Suppose you want to write a programme to interactively compute the response of a low-pass filter, with the number of poles not fixed in advance. There are at least two possible solutions, inductance-input and capacitance-input, with the added unknown of the number of poles (Fig 10.1).

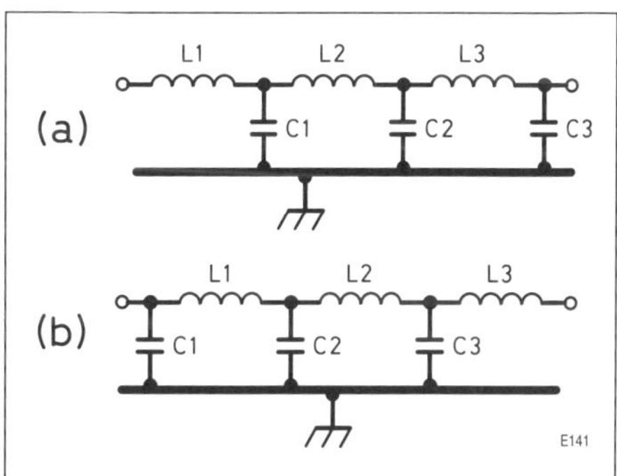


Fig 10.1

So, how would you manage this in a strictly procedural language? There are not many choices. You have to pose a series of questions to the user of the programme, to understand what he has in mind, something like this:

```
C:\>dofilter
Please type 1 if inductance-input, 2 if capacitance-input? 1
Now please type the number of poles: 4
Ok, now type the value of L1: 100e-6
Ok, now type the value of C1: 560e-12
...... etc. etc. etc...............
```

This would of course work, but what if you change your mind after having entered all of the values, and you want now to model a capacitance-input filter? You must redo everything from the beginning for a different number of poles.

On the other hand, a very simple event-driven Windows programme to accomplish the same task could look like that shown in Fig 10.2.

The user enters the data in any order, and only when he presses the 'Compute response' button is all the data analysed and the response computed. And don't be misled by the fact that the above panel has only four input fields both for capacitance and inductance. When the 'Number of poles' input field receives a valid value, more (or less) fields where to specify capacitance and inductance values will be displayed, with the panel changing size to accommodate them.

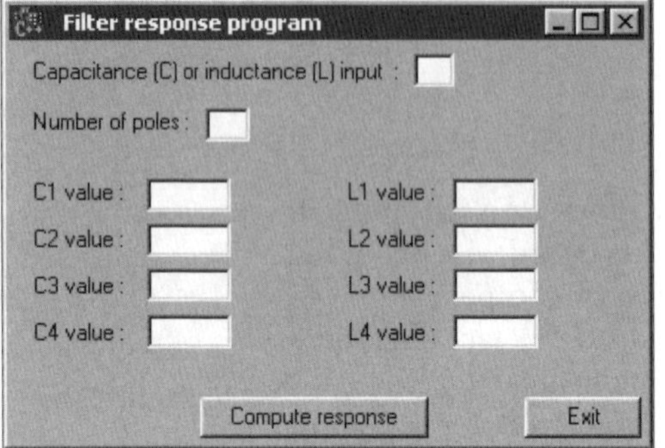


**Fig 10.2**

What happens behind the scenes? Each time a field receives a value or a button is pressed, Windows notifies the fact to the programme, which must then react accordingly, eg by storing the value of the component or changing the computation algorithm that will be afterwards used, or showing a different number of input fields on the panel. In other words, the programme must react to the *events* caused by user actions and duly reported by Windows. These events can be generated in any order, and the programme must be prepared for this, possibly generating error messages if the input is invalid (eg missing number of poles or the value of a component). This is maybe the greatest change of attitude the Windows programmer has to face when switching from a command line environment. He is no more in control of what happens – the user is.

To better explain all of the above, and to render justice to the title of this chapter, we will now examine in detail a simple Windows programme, whose purpose is to read digitised audio data from the sound card, apply a comb filter to it, and play back the filtered audio from the speakers. A comb filter, in case you don't know, is just a series of narrow notches placed at the multiples of a given frequency. It derives its name from the fact that its spectral response looks like an ordinary comb. Its main use is to filter out the hum generated by the mains and its harmonics. To make it useful also on the other side of the Atlantic, the fundamental frequency can be chosen to be 50 or 60Hz. As it seems to work rather well, I gave it the name *HUMID*, which is an acronym of 'HUM Instant Destroyer'. Please don't blame me for this choice of name! Its panel looks like Fig 10.3.

Special attention will be given to the sound card interfacing portion of the programme, so that it can be taken as a model for whatever programs one could wish to write

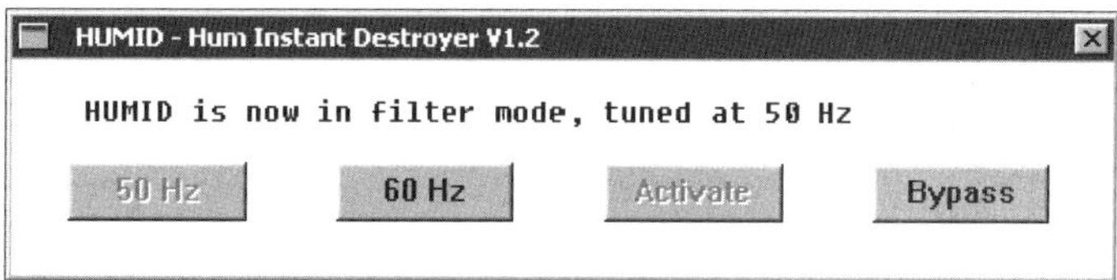


Fig 10.3. HUMID

to process audio, for a amateur radio related (or otherwise) application. The comb filter implementation is in a separate subroutine, and the algorithm is modelled after a programme of Paul Nicholson [1] with his permission. The entire listing of HUMID is in Table 10.14 at the end of this chapter, and we will examine it step by step.

I suggest the reader gets hold of the free Borland C/C++ compiler, which can be downloaded from reference [2]. When you are on that page, select the third choice in the table, the one that says 'Compiler'. You will have to agree to the licence statement and to register, but then the download is free of charge. With this compiler you will be able to compile HUMID and to test modifications you can make to it, or to build your own Windows application. Follow the instructions in the package regarding installing; it is indeed a very simple process.

# HUMID

I said before that everything is governed by events. I should have more correctly said that *messages* are the key pivot at the heart of Windows. When an event happens, Windows sends a message to the programme that 'has the focus', which is an expression that indicates the window (programmes interact with the users through panels called *windows*) that, at that point in time, will receive all the mouse and keyboard input. This generally is the window which is in front of all the others (but there are exceptions). The programme examines the message and decides whether it has some actions to perform to handle it, or not. In other words, the programme sits in an idle loop, waiting for messages from Windows. Once it has processed a message, it just waits for the next. This loop is called *the window procedure* in practically all books devoted to the subject of Windows programming. The programme, as part of its initialisation, sets up that window procedure, and passes a pointer to it to Windows.

Let's see how this is done in HUMID. The beginning of the programme is shown in Table 10.1.

Nothing strange here, we have declared some variables we will need afterwards. The structure array `button` contains some data (the style, the caption and whether it is enabled) that will be used to create the four pushbuttons present in the HUMID window. Generally speaking, all the variable types in uppercase are defined inside the header `windows.h` which in turn is included by `humid.h`. Don't be too worried about their exact meaning just now – when you are more proficient in the art of Windows programming, you can look them up in the Windows documentation. Just a few additional words: `srate` is the sampling rate that the sound card will

**Table 10.1**

```
#include "humid.h"
struct
{
    long style ;
    char *text ;
    BOOL enabled;
}   button[] =
    {
        BS_PUSHBUTTON,        "50 Hz",    FALSE,
        BS_PUSHBUTTON,        "60 Hz",    TRUE,
        BS_PUSHBUTTON,        "Activate", FALSE,
        BS_PUSHBUTTON,        "Bypass",   TRUE
    };

BOOL        Hz50 = TRUE, Active = TRUE;
int         srate = RATE;
HWND        hwnd;
HANDLE      BufferThread, ProcessThread, InBufferReady, OutBufferReady;
HWAVEIN     hin;
HWAVEOUT    hout;
WAVEHDR     inhdr[NINBUF], outhdr[NOUTBUF];
RECT        rect, screen ;
```

be set to. `BufferThread` and `ProcessThread` are the *handles* (fancy word that has more or less the meaning of 'pointer') of the two additional *threads* (besides the main programme) that compose HUMID.

I hear many of you saying, "Threads? What on earth do you mean by that?" You are quite right. I haven't talked about threads yet. Explaining them fully is a bit beyond the purpose of this chapter, but in a nutshell a thread is an indivisible unit of execution, a chunk of work that the CPU is assigned to chew upon, until it is somehow interrupted and assigned a new thread. Windows itself is composed of threads, and so are user programs. It is not uncommon to find hundreds, even thousands, of threads present in your PC at some point in time. The advantage of threads is that they can be interrupted (and resumed later on), so that the CPU can work on a higher priority one if need be. In the case of HUMID we have two of them, plus the main programme, also a thread. One to manage the re-queuing of the input audio buffers to the sound subsystem (more on this later) and the other to perform an analogue function for the output buffers just after the signal processing functions of the comb filter has been done. Using threads has the advantage of higher user interface responsiveness. In other words, what would happen without threads is that, for example, if you pressed a button while the programme is engaged in a lengthy computation, your action would not be recognised and acknowledged until the end of the computation, causing a sluggish response by the programme.

Back to the HUMID code (Table 10.2). `WinMain` takes the place of the `main` routine of a classical C programme. Its parameters are almost always as declared above – just use this format and you will be fine. With the call to the Windows API `GetWindowRect` ('API' is an acronym of 'Application

**Table 10.2**

```
int WINAPI WinMain (HINSTANCE hInstance, HINSTANCE hPrevInstance,
                    PSTR szCmdLine, int iCmdShow)
{
    static char szAppName[] = "Humid" ;
    MSG            msg ;
    WNDCLASSEX  wndclass ;

    GetWindowRect(GetDesktopWindow(), &screen);

    rect.left = screen.right/2 - 250;
    rect.top  = screen.bottom/2 - 60;
    rect.right = rect.left + 500;
    rect.bottom = rect.top + 120;

    wndclass.cbSize        = sizeof (wndclass) ;
    wndclass.style         = CS_HREDRAW | CS_VREDRAW ;
    wndclass.lpfnWndProc   = WndProc ;
    wndclass.cbClsExtra    = 0 ;
    wndclass.cbWndExtra    = 0 ;
    wndclass.hInstance     = hInstance ;
    wndclass.hIcon         = LoadIcon (NULL, IDI_APPLICATION) ;
    wndclass.hCursor       = LoadCursor (NULL, IDC_HAND) ;
    wndclass.hbrBackground = (HBRUSH) GetStockObject (WHITE_BRUSH) ;
    wndclass.lpszMenuName  = NULL ;
    wndclass.lpszClassName = szAppName ;
    wndclass.hIconSm       = LoadIcon (NULL, IDI_APPLICATION) ;
```

Programme Interface', all the functions that Windows makes available to user programs to perform a variety of tasks – there are more than 1000 of them), we are asking what the screen size is to place our window just in the centre of the desktop. The following instructions set some characteristics of the `wndclass` structure, which represents our window. I won't go through each one, as you can easily find what they do by perusing a Windows API reference book, which is a highly suggested companion to be kept near the PC keyboard. One instruction to be noted is:

```
wndclass.lpfnWndProc = WndProc;
```

Here we set inside the `wndclass` structure the pointer to our window procedure, our message processing loop. Talking of reference material, a book you cannot do without is the excellent *Programming Windows 98* by Charles Petzold, published by Microsoft Press. Everything not explained here for space reasons is covered in that book, in more than 1000 pages! If you are serious about writing Windows programs, get hold of it.

Let's continue with the HUMID code (Table 10.3). We are building here the foundations of our programme. First, we register with Windows our window procedure class `wndclass`, then we create the main window of the programme, specifying its caption, the style, where it must positioned on the desktop and its size. Then we show it by calling `ShowWindow`. Note that this API shows only the Windows-managed part of the panel, ie the

**Table 10.3**

```
RegisterClassEx (&wndclass) ;
hwnd = CreateWindow (szAppName, "   HUMID - Hum Instant Destroyer V1.2",
                     WS_CAPTION | WS_SYSMENU,
                     rect.left, rect.top,
                     500, 120,
                     NULL, NULL, hInstance, NULL) ;

ShowWindow (hwnd, iCmdShow) ;
UpdateWindow (hwnd) ;
```

borders, the title, and the small icons at the top left and top right. The actual content of the window, the so-called *client area*, is left to the responsibility of our programme, and this is done by calling UpdateWindow. The effect of this call is just to generate a message to our window procedure, telling it that the client area of the main window must be refreshed. This is a very important point to keep in mind. Our programme must always be ready to regenerate the entire contents of the window upon request by the operating system. Don't overlook this point. Whatever change you make to the window contents, you must be able to do it again, should it be needed. And the need could be caused, by example, by your window being hidden by another one, and then brought again in front. Windows doesn't save what you were showing. When your window becomes again visible, it just asks you to redraw it. This is another point where the apprentice Windows wizard often fails. The implication of all this is that you must keep track of the status of what you are showing. When we declared the four pushbuttons at the beginning of the programme, we used a structure that has a member named enabled, which is a Boolean variable. This variable holds the current 'pushability' status of its associated button. If I push the 50Hz button, then it must become disabled, as there is no point in pressing it once more. Instead, the 60Hz button must now be enabled. By storing the status of the buttons, it is then possible to redraw them, when Windows asks, in a way showing their respective enabling status.

Let's go on (Table 10.4). Finally we have found the main processing loop, the heart of our programme. The GetMessage API doesn't return to the caller until Windows has a message for us. When there is one, we call TranslateMessage. The TranslateMessage function translates virtual-key messages into character messages. The character messages are posted to the calling thread's message queue, to be read the next time the thread calls the GetMessage or PeekMessage function. To actually send the message to our window procedure, we use DispatchMessage. The DispatchMessage function dispatches a message to a window procedure. It is typically used to dispatch a message retrieved by the GetMessage function. This last sentence has been copied verbatim from the API help. If all this looks unnecessarily complicated and involved, well, it is. But then you haven't seen anything yet, as far as Windows is concerned.

Note the return msg.wParam instruction. This is where our programme

**Table 10.4**

```
    while (GetMessage (&msg, NULL, 0, 0))
    {
        TranslateMessage (&msg) ;
        DispatchMessage (&msg) ;
    }

    return msg.wParam ;
```

ends. And we reach this instruction when the `GetMessage` API returns a return code of zero. This is the method used in Windows to terminate a programme. But now we have reached the point where things get done, the window procedure (Table 10.5). Basically it is a giant `switch case` construct, where the identifier of the message just received from Windows is examined and acted upon.

There are a lot of things to explain here and a problem: if I go too much in detail, then the outcome will be a replica of Petzold's book. Too little detail, and you will feel dissatisfied, not having grasped enough of the subject to justify the use of your time reading this chapter. I will try to stay in the middle ground, but please bear with me if I don't always succeed!

We will examine the declarations at the beginning of this routine while we use them. Note that the entire procedure is a single `switch case` construct, and the code above reports the first `case`, the `WM_CREATE`. Windows sends a `WM_CREATE` message to our procedure when, logically enough, our window is being created, in order to allow us to perform one-time initialisation tasks like, in our case, setting up the comb filter, opening the sound card etc.

The first thing we have to do is to know how large the characters on the screen are! This may sound odd, but it isn't. Windows permits the use of various character sizes with different fonts, and we want to size our buttons so that their labels do not spill outside the button borders. Without entering into too much detail, first we obtain a so-called *device context*, which is a handle to the virtual painting area on the screen. We do this with the `GetDC` call. Then, using a couple of APIs, we set the two variables `cxChar` and `cyChar` that indicate the horizontal and vertical sizes of the characters used. Now is the time to actually create the four small windows that represent our four buttons. Remember, in Windows almost everything is a window (pardon the pun), so the buttons are created with the `CreateWindow` API call, using the `button` structure declared at the beginning of the programme, and the character sizes just found to compute a correct button dimensioning.

Then we try to initialise the comb filter – in case of failure we display a "We are sorry but . . ." type of message, and we also tell Windows to put into the message queue for our programme a `WM_QUIT` message, which will be retrieved by the `GetMessage` API, causing it to return with a return code of 0 (zero), terminating the programme. We will examine the `SetupComb` function later on. Same logic for the two functions `SoundOpenInput` and `SoundOpenOutput`, whose description is also

**Table 10.5**

```
LRESULT CALLBACK WndProc (HWND hwnd, UINT iMsg, WPARAM wParam, LPARAM lParam)
    {
    static char   msg[128] = "HUMID is now in filter mode, tuned at 50 Hz";
    static HWND   hwndButton[NUM] ;
    static int    cxChar, cyChar ;
    HDC           hdc ;
    PAINTSTRUCT   ps ;
    int           i, k;
    TEXTMETRIC    tm ;
    long          style;
    DWORD         thid1, thid2;

    switch (iMsg)
    {
        case WM_CREATE :
            hdc = GetDC (hwnd) ;
            SelectObject (hdc, GetStockObject (SYSTEM_FIXED_FONT)) ;
            GetTextMetrics (hdc, &tm) ;
            cxChar = tm.tmAveCharWidth ;
            cyChar = tm.tmHeight + tm.tmExternalLeading ;
            ReleaseDC (hwnd, hdc) ;

            for (i = 0 ; i < NUM ; i++)
            {
                style = WS_CHILD | WS_VISIBLE | button[i].style;
                if(!button[i].enabled) style |= WS_DISABLED;

                hwndButton[i] = CreateWindow ("button", button[i].text,
                                style,
                                cxChar * 15 * i + 25, cyChar * 3,
                                10 * cxChar, 7 * cyChar / 4,
                                hwnd, (HMENU) i,
                                ((LPCREATESTRUCT) lParam) -> hInstance,
                                                        NULL) ;
            }

            if(!SetupComb())    // initialize the comb filter
            {
                fatal("Unable to allocate filter buffers");
                PostQuitMessage (1) ;
            }


            if(!SoundOpenInput() || !SoundOpenOutput())
                    PostQuitMessage (1) ;

            InBufferReady  = CreateEvent(NULL, FALSE, FALSE, NULL);
            OutBufferReady = CreateEvent(NULL, FALSE, FALSE, NULL);

            BufferThread   = CreateThread(NULL, 0, ReQueBuffers,
                                        0, 0, &thid1);
            ProcessThread  = CreateThread(NULL, 0, OutputAudio,
                                        0, 0, &thid2);
            if(NULL == BufferThread || NULL == ProcessThread)
```

**Table 10.5 (continued)**

```
        {
            fatal("Unable to create processing threads");
            PostQuitMessage (1) ;
        }

            SetThreadPriority(BufferThread,  THREAD_PRIORITY_HIGHEST);
            SetThreadPriority(ProcessThread, THREAD_PRIORITY_HIGHEST);

// if everything ok so far, then tell MMAPI to start recording
// but only when we will have plenty of free output buffers

            while(TRUE)
            { for(k=0; k<NOUTBUF; k++)
                if(outhdr[k].dwUser != 0) break;

                if(k>=NOUTBUF/2) break;
            }
            waveInStart(hin);

            return 0;
```

postponed. Next we create the two threads mentioned previously, namely `ProcessThread` and `BufferThread`, together with the two events that will be used to awake them from the idle status when we will need their services, ie `InBufferReady` and `OutBufferReady`. Here too we make sure that everything went well, aborting if not. As a final touch, we modify the priority of the two threads to make them ahead in the Windows scheduling queue. This helps to prevent interruptions and choppiness in the sound output.

We are now at the end of the initialisation process; we just have to crank up the sound subsystem, which is composed of the playback and the input parts, running independently. The playback part starts automatically as soon as it has been initialised (we will do it in the `SoundStartOutput` routine, described later), while the input part waits for a specific command. We take advantage of this in a way to not cause buffer starvation, which is what happens when the sound card has just digitised a new chunk of audio, but there are no free buffers to store it in, or the playback portion of the sound card has finished playing back a buffer and is asking for a new one, but there are none ready. To avoid that, the audio buffers are marked using the `dwUser` field, a field not used by Windows, but left to the user. We put in this field the value of 0 (zero) to indicate that this particular buffer is free, a value of 1 to indicate that it has been filled with audio contents, and a value of −1 to indicate that it has been put into the playback queue, ready for when Windows asks for a new output buffer.

Returning to our code, we enter into a `while` loop which will be exited only when at least half of the output buffer is free, having been played back. The output buffers are primed with a content of 0 (zero) which corresponds to silence. So, while the second half of the output buffers is being played

**Table 10.6**

```
case WM_COMMAND :
    i = LOWORD (wParam);
    k = (i > 1) ? (5 - i) : (1 - i);

    button[i].enabled = FALSE;
    button[k].enabled = TRUE;

    EnableWindow(hwndButton[i], FALSE);
    EnableWindow(hwndButton[k], TRUE);

    Hz50   = !button[0].enabled;
    Active = !button[2].enabled;

    wsprintf(msg, "HUMID is now in %s mode", Active ? "filter" :
                                             "bypass");
    if(Active)
    { strcat(msg, ", tuned at ");
      strcat(msg, Hz50 ? "50 Hz" : "60 Hz");
    }
    else strcat(msg, "                          ");

    hdc = GetDC (hwnd) ;
    SelectObject (hdc, GetStockObject (SYSTEM_FIXED_FONT)) ;
    TextOut (hdc, 4 * cxChar, cyChar, msg, strlen(msg)) ;
    ReleaseDC (hwnd, hdc) ;

    return 0;
```

back, we fill the first half with new audio data. To do this, we explicitly start the input part of the sound subsystem using the API `waveInStart` when we exit from that `while` loop. Having done this, the initialisation is completed, and we return with a return code of zero to Windows. This ends the management of the `WM_CREATE` case of the `switch` clause.

The next message to manage is the `WM_COMMAND` one (Table 10.6). Windows sends us this message when the user is interacting with our programme, in this case by clicking on a pushbutton with the mouse.

Sending us a `WM_COMMAND` message is not enough, of course. We need to know at the minimum which button has been clicked. Looking carefully at the parameters passed to our window procedure, you will see `wParam` and `lParam`. This latter contains the handle of the child window clicked. I told you that everything is a window, buttons included. The windows contained in the main window are aptly named *child windows*, as they inherit some characteristics from the parent window, and are automatically closed when the parent window closes. The child window handle is the value that Windows returns from the `CreateWindow` call. But in this programme we won't make use of this handle, as there is more useful information passed into the `wParam` parameter. This is a 32-bit value, split into the low- and the high-word parts.

The high-word part carries the so-called *notification code*, which tells what has been done with the button. There is not much else that can be done with a button, apart from clicking it, so this word will contain always the value of 0 (zero), corresponding to the manifest constant `BN_CLICKED`, defined by the `windows.h` header. The low word has the ID of the specific button that has been pressed. This ID is the value passed to `CreateWindow` when the child window was created, the 9th argument of the call, the variable `i`, which was a zero-based index, ranging from 0 (zero) to 3. We use this ID to perform the correct function for the button clicked, and also to modify both the logical and the physical status of that button. What I mean with this is that, once pressed, the button must be greyed out to indicate that it cannot be clicked a second time, while the other one must now be enabled. For this we use one of those dirty tricks that non-orthodox programmers like me love:

```
k = (i > 1) ? (5 - i) : (1 - i);
```

In the above statement we compute the index k of the button paired with that pressed: i can assume the values 0, 1, 2, 3 – what we want is that if i = 0, then k = 1 and vice versa, if i = 2 then k = 3 and vice versa. Instead of using a boring sequence of `if` statements, this can be elegantly accomplished with the single statement above. Now we can use the indexes i and k to modify the status of our buttons, and to set a couple of Boolean variables, namely `Hz50` and `Active` which will be used later on, when actually doing the filtering. We are almost finished with the `WM_COMMAND` case – we just put on the screen a message telling the user what the present status of the filter is. The main API used for this is `TextOut` preceded by the selection of the system fixed font used for writing the string. You can look up these APIs in the reference book.

There is an important point to remark here, in case you missed it. We are in the message-processing loop of our programme, where responsiveness to user input is of utmost importance. We cannot afford to perform lengthy processes here. What we in the end do when the user clicks a button is just to set a couple of Boolean variables, which will then be examined by asynchronous processes (threads). Keep this in mind, follow this rule, and your first step towards user-friendliness is done.

Now for the `WM_PAINT` and `WM_CLOSE` cases (Table 10.7). The `WM_PAINT` is the message that Windows sends us when there is a need to redraw the client area of the window. In the case of this programme it is quite simple, as there is just one thing that we have to refresh on the screen upon Window's request, ie that string showing the filter status, which was dynamically built by us in response to the `WM_COMMAND` message. The four buttons are child windows that do not have a client area, so Windows takes care itself of refreshing them. So what we have to do is just to acquire a device context handle (our virtual painting canvas) by calling the `BeginPaint` API, reuse the same `TextOut` API that we used in the `WM_COMMAND` case, and finally release the device context handle with `EndPaint`. That's all. It would have been more complicated had ours been a graphical programme, maybe with layer support.

**Table 10.7**

```
        case WM_PAINT :

            hdc = BeginPaint (hwnd, &ps) ;
            SelectObject (hdc, GetStockObject (SYSTEM_FIXED_FONT)) ;
            TextOut (hdc, 4 * cxChar, cyChar, msg, strlen(msg)) ;
            EndPaint (hwnd, &ps) ;

            return 0 ;

        case WM_CLOSE :

            PostQuitMessage (0) ;
            return 0 ;
```

The final `case` is `WM_CLOSE`. This message is sent by Windows typically when the user clicks on the small x-shaped icon at the upper right corner of the window. Windows politely asks us to terminate, please, giving us the chance to perform termination tasks, if need be, like releasing resources, closing files etc. When we are ready to end the instance of our existence in the current Windows session, we communicate this decision of ours by using the `PostQuitMessage` API. The `PostQuitMessage` function posts a `WM_QUIT` message to the thread's message queue and returns immediately; the function simply indicates to the system that the thread is requesting to quit at some time in the future. As previously explained, a `WM_QUIT` message causes the `GetMessage` API to return a 0 (zero) value, thus exiting the `while` clause and bringing the life of the programme to an end. Note that, if you do not call the `PostQuitMessage` function, your programme does not terminate, regardless of how vehemently the user is clicking on that x-shaped icon, but don't then complain if yours won't be deemed an user-friendly programme!

We have finished examining the `switch` construct, which has no *default case*. So, when we receive from Windows a message not explicitly managed by us, we just pass it back to the default procedure of Windows that takes care of unmanaged messages:

```
        return DefWindowProc (hwnd, iMsg, wParam, lParam) ;
```

And with this we have completed our window procedure, where we handle the messages originated by Windows, typically in response to user actions. What remains is the sound card initialisation and handling, and the actual comb filter code. We examine them in the next sections.

### The sound card

You may recall that in the `WM_CREATE` case we encountered the two calls `SoundOpenInput` and `SoundOpenOutput` which weren't described in detail at that moment. We will do it now. `SoundOpenInput` does nothing more than calling `SoundStartInput` and checking its return code, alerting the user if something went wrong. In computer programming jargon, this is called a *wrapper function*. You will likely find this word again, should you further pursue your Windows programming adventure.

Before delving into the sound card code, maybe a few words on the general architecture of sound recording and playing back are in order. Windows has two major systems for audio management. The first is called *DirectX*. You surely have already heard of it already. DirectX is a component (optional in some versions of Windows) meant for generally managing multimedia contents such as audio, video, animations, images and the like. The second is the *WMME*, which stands for 'Windows Multimedia Extensions', and it is standard in all versions (as far as I know) of Windows, from '95 up.

I generally use this latter system. Why? Well, DirectX perhaps has a lower latency time (the time elapsed between the moment you give a buffer to Windows to play back, and the moment you actually hear it from the PC speakers), but has some drawbacks. First, you (the developer) do not have the absolute guarantee that DirectX is installed on the user's PC, so you must provide an installation procedure for it, just in case. Second, Microsoft from time to time releases new versions of DirectX, which do not always maintain a full compatibility with the older versions. In my modest opinion these two drawbacks outweigh the small advantage in terms of latency time. Just a personal opinion, of course. I tested DirectX, but wasn't overly impressed by it. To be honest, it does have another advantage, ie with DirectX you can share the sound card with other applications running simultaneously on the PC, but with the advent of the AC97 standard (more on this would be beyond the scope of this chapter) you can do the same with WMME too.

So, having chosen WMME, how are they used? Conceptually, it is very simple. You open the input (or output) device, allocate a few buffers, pass them to the WMME, and you are done. When a new input buffer is ready, or there is a need for another output buffer to be played back, Windows interrupts asynchronously the flow of the programme in one of the various ways that you have specified during the open routine. The simplest one, which I have used here, is the so-called *Callback Procedure*, which is a routine called directly by the inner Windows thread belonging to WMME. A word of caution here: this routine is executed as part of the system WMME thread, so you are a bit restricted as to what you can do here. The safest way is to copy the data, set flags if need be, and just return to the caller.

Let's now look into the `SoundStartInput` routine (Table 10.8). The first thing done here is to be sure to have at least one MCI device ('MCI' stands for 'Media Control Interface', in other words we check if a sound card is present and correctly installed), using the `waveInGetNumDevs` API, then we set a few fields of the `WAVEFORMATEX` structure, used subsequently by the open function. Specifically, we set the number of channels, 1 in this case as we are not interested in stereo, the format of the data which is the standard PCM (pulse coded modulation, the normal uncompressed WAV format) and the sampling rate, which, for this application, is 44100 samples per second. We use the maximum sampling rate supported by the sound card, as the tracking algorithm of the comb filter works better the more

**Table 10.8**

```c
MMRESULT SoundStartInput(void)
{
    int i;
    MMRESULT ierr;
    WAVEFORMATEX    wf;

// first check that we have at least *ONE* MCI device

    nDevIn = waveInGetNumDevs() - 1;
    if (nDevIn < 0) return 1;

    ZeroMemory(&wf,sizeof(wf));     // clear the structure with 0
    wf.wFormatTag = WAVE_FORMAT_PCM;
    wf.nChannels = 1;
    wf.nSamplesPerSec = srate;
    wf.wBitsPerSample = (WORD) (8 * nByteIn);
    wf.nBlockAlign = (WORD)((wf.nChannels * wf.wBitsPerSample) >> 3);
    wf.nAvgBytesPerSec = wf.nSamplesPerSec * wf.nBlockAlign;
    wf.cbSize = 0;
    ierr = waveInOpen((LPHWAVEIN)&hin, WAVE_MAPPER, (LPWAVEFORMATEX)&wf,
                        (DWORD)SoundProcIn, 0L, CALLBACK_FUNCTION);
```

frequently it receives data to act upon. We set also the number of bits for each sample which, for PCM, is 16. Each sample ranges from −32768 to 32767, the value of 0 (zero) representing silence.

During audio manipulation it's usually more convenient to work with floats to minimise rounding and ranging errors, and this can be done by converting shorts (the 16-bit values) to floats and back again just before playing them back. However, this is not necessary in Humid as there are no computations (like FFTs) that could benefit from an extended dynamic range, so we will deal directly with short integers, as passed from the WMME.

It's now time to actually open the input stream of audio. The API used for this is waveInOpen, to which we pass, amongst other parameters, the address of the callback procedure SoundProcIn that WMME will call each time a new input buffer is filled with audio data. Good programming habits dictate that a check is done on the return code from waveInOpen, to see whether everything went well. I don't repeat here the details of the code (you can see it in Table 10.14) but in a nutshell we simply check for various error codes, displaying the appropriate error message should one of those happen. If everything is fine, it's time to give some buffers to the input audio subsystem. Those buffers are allocated statically in this programme, but other applications may prefer to allocate them dynamically, perhaps of size and number determined at run time, based on the sampling rate specified by the user and generally by other requirements. When buffers are given to the WMME, they must be *prepared*, more technical jargon which means that Windows must prime their headers with some initial information. The code is shown in Table 10.9.

What we do is simply to initialise the user fields of the headers with pointers

**Table 10.9**

```
    for(i=0; i<NINBUF && ierr == MMSYSERR_NOERROR; i++)
    {
        ZeroMemory(&inhdr[i],sizeof(inhdr[i]));
        ZeroMemory(inbuf[i],BUFSIZE * nByteIn);
        inhdr[i].lpData = (char *) inbuf[i];
        inhdr[i].dwBufferLength = BUFSIZE * nByteIn;
                    /* The Length must be in bytes*/
        inhdr[i].dwBytesRecorded = 0;
        inhdr[i].dwUser = 0;
        inhdr[i].dwFlags = 0;
        inhdr[i].dwLoops = 0;
        inhdr[i].lpNext = NULL;
        inhdr[i].reserved = 0;
//first prepare the waveheaders
        ierr = waveInPrepareHeader(hin, &inhdr[i], sizeof(WAVEHDR));
//second add the buffers to the MMAPI queue
        ierr += waveInAddBuffer(hin, &inhdr[i], sizeof(WAVEHDR));
    }
```

to the actual buffer space and to set a few flags, amongst which is the dwUser field mentioned before, used to keep track of current status of that particular buffer. Nothing fancy here, just normal administration. Windows also has its fields to prime, as said before, and we let it do that by calling the waveInPrepareHeader API. Last step is to actually pass those buffers to the WMME with the waveInAddBuffer API. This call should return 0 (zero) if all is well, so to make things simpler, we just add the return codes inside the loop for each buffer and we test at the exit of the loop if the sum is zero. We are not interested in knowing which (if any) of the calls failed. And practically this call doesn't ever fail. Now everything is ready; we just have to start up the input side of audio management, and this is done by calling the function waveInStart which you may recall we have already encountered in the WM_CREATE case of our window procedure.

The output part of audio handling is practically the same as the input part, and it is implemented in the SoundStartOutput routine. Of course waveInOpen will become waveOutOpen and so on. But one difference is to be remarked. Here we stop the production of output using waveOutPause just after the opening, and we resume it later on, after having passed all the output buffers to the WMME, by using the waveOutRestart API. This because the playback doesn't need to be started up manually – it starts automatically as soon as the waveOutOpen is issued.

Fine, the engine is started and buffers are happily flowing in and out. How do we manage them? If you remember, we specified our desire to use callback functions for this purpose. These two functions are SoundProcIn and SoundProcOut. Let's look at them (Table 10.10).

The routine is really simple. We test why it has been called: whether during open/close, or for a new buffer ready. In this latter case it in turn calls ProcessBuffer, which is shown in Table 10.11.

Here the newly arrived buffer, pointed to by the field lpData of the

**Table 10.10**

```
VOID CALLBACK SoundProcIn(HWAVEIN hwi, UINT iMsg, DWORD dwInst,
                          DWORD dwParam1, DWORD dwParam2)
{
    switch(iMsg)
    {
    case WIM_DATA:
        if(!InClosing)
        {
            ProcessBuffer((LPWAVEHDR) dwParam1);
        }
        break;

    case WIM_OPEN:  // Message sent when input device is opened
        sndInOpen = TRUE;
        break;

    case WIM_CLOSE: // Message sent when input device is closed
        sndInOpen = FALSE;
        break;
    }
}
```

**Table 10.11**

```
void ProcessBuffer(LPWAVEHDR ibuf)
{

    memcpy(tmpbuf, ibuf->lpData, BUFSIZE * sizeof(short));
    SetEvent(InBufferReady);
    Filter();
    SetEvent(OutBufferReady);
}
```

structure iBuf (the buffer header), is copied into a working buffer, then an event is set to awake the ReQueBuffers thread, whose task is to put that buffer again in the input queue, ready for new audio contents. Why do we need to have a thread dedicated to this? Couldn't have we done that directly in the ProcessBuffer routine? The answer is, unfortunately, no. If you remember, I told you that the callback procedure is executed as part of the WMME thread, and some APIs are not permitted here. waveInAddBuffer is one of them, so we have to resort to raise a flag that will ask for the services of a different thread to put again that buffer into the input queue. A word of caution here: Windows 95/98 is rather tolerant. It lets you issue that call inside the callback procedure without complaining and the programme will even work. But don't try to execute such a programme under Windows 2000 or Windows XP – failure is guaranteed.

Besides raising the InBufferReady flag, we do a couple of other things. The actual filtering is done here by calling the Filter function. This seems to contradict what I said earlier, that a minimal amount of work is to be done inside the callback procedures. As a matter of fact, the comb filter doesn't consume a lot of processing power. I measured roughly 10% of a

**Table 10.12**

```
void Filter(void)
{
    if(NULL == (newOutBuffer = GetFreeOutBuffer())) return;
    if(Active) ApplyComb(tmpbuf, Hz50 ? 50. : 60.);
    memcpy(newOutBuffer->lpData, tmpbuf, BUFSIZE * sizeof(short));
}
```

750MHz Athlon CPU, so it's safe to put the filtering code executed as part of the WMME thread. Things would be different in case of CPU-intensive processing code. In such a case, the safest bet would be to dedicate a specific thread to that task, again using a `SetEvent` call to signal it that it has work to be done. This is the technique I used, for example, in the programme Spectran that I distribute on the Internet. The `Filter` function is very simple (Table 10.12).

Here we first get a free output buffer from the free buffers pool by calling the `GetFreeOutBuffer` function. Then the `ApplyComb` (the real filtering code) function is called but *only* if the `Active` Boolean variable is set. You may recall that we set or reset this variable in response to the user action of clicking one of the two buttons ('Bypass' or 'Activate') on the main window of Humid. The filtering is done in-place, which means that audio data are passed in `tmpbuf`, and from the same buffer they are retrieved. So we can safely copy that `tmpbuf` into the previously acquired free output buffer, without caring if the comb filter has actually been applied. Pressing the 'Bypass' button has the effect of copying the input buffer to output, unchanged, which is exactly what we wanted to do.

Exiting from `Filter` we go back to the last statement of the `ProcessBuffer` function, called by the input callback procedure, where we just set a flag for the output queuing thread `OutputAudio` which simply scans the output buffers pool and queues to the output part of the WMME all of the buffers that have the `dwUser` flag (remember that?) set to 1, which means "filled, but not queued yet". The flag is changed to −1 to indicate that the queuing has been now done. Again, we need to resort to a separate thread for the same reasons mentioned above.

Turning now to the output callback procedure `SoundProcOut`, what we do here is really simple (Table 10.13).

This procedure is called when an output buffer has been played back, and it is passed the address of that buffer. All we have to do is to set the `dwUser` flag of the buffer to 0 (zero), to return that buffer to the free output buffers pool. That's all. We don't need to feed here the output engine with new output buffers – this task, as we have already seen, is done just after the comb filtering. As soon a new output buffer is filled with new audio data, it is queued for playback. In this way we minimise the risk of buffer starvation.

This concludes the description of sound card management in Humid. In the next section we examine briefly, as delving into DSP algorithms is not the purpose of this chapter, the comb filter code that performs the actual hum removal from the audio signal.

**203**

**Table 10.13**

```
VOID CALLBACK SoundProcOut(HWAVEOUT hwo, UINT iMsg, DWORD dwInst,
                               DWORD dwParam1, DWORD dwParam2)
{
    switch(iMsg)
    {
      case WOM_DONE:
        if (!OutClosing)
        {
            ((LPWAVEHDR) dwParam1)->dwUser = 0;
        }
        break;

      case WOM_OPEN:  // Message sent when output device is opened
        sndOutOpen = TRUE;
        break;

      case WOM_CLOSE: // Message sent when output device is closed
        sndOutOpen = FALSE;
        break;

    }
}
```

### The comb filter

The comb filter belongs to a category of digital filters that are easy to implement in the time domain, consume few resources and are very effective in doing what they are designed for. They belong to the class of FIR (finite impulse response) filters, which is jargon meaning that to compute the $N$th filtered output sample you combine somehow the $N$th input sample and some of the previous samples, going back as long as the order of your filter dictates. They are called 'finite impulse response' because there is the guarantee that when the input goes to zero, so will do the output signal in a finite time, which depends on the order of the filter. This is to be contrasted with the IIR (infinite impulse response) filters, which use also some of the previous outputs to compute the current one. For IIR filters the guarantee of having eventually zero output with zero input does not hold, given the method of computation of the current output.

Paul Nicholson, the author of the original routine from which the comb filter of Humid has been adapted, has written the following notes about its filter.

"This filter copies the sound card input to the output, tenaciously following the line frequency hum with a deep notch. When properly set up, the harmonic components of foreground hum should almost all the time be completely inaudible at 16-bits dynamic range.

"Comb filtering is very effective at removing the harmonic components of line frequency interference. By subtracting a delayed version of the signal from the original, any incoming frequencies which are periodic in the delay time are completely notched out. The frequency width of the notches is fairly wide and produces a noticeable drainpipe effect on the sound. By combining the output of multiple delay stages, this effect is considerably reduced and a much more natural sound is produced.

"Sharpness of the multiple stage notch necessitates accurate tracking of the varying line frequency (typically wandering by up to ±0.5%) to ensure that the line harmonics remain in the deepest point of the notches. This can be done by using the delay buffer and summing function to generate a three-point period-gram, which is used to steer the delay time towards that which gives maximum output from the summing function, and therefore the deepest notch.

"The depth of the notches and the tightness of the period lock enable the harmonic components of foreground hum to be removed almost completely. The programme uses linear interpolation when drawing samples from the delay buffer, which improves the accuracy of the notch for those delay times.

"Comb filtering works by subtracting a delayed version of the input signal from the original. Signals which are periodic in the delay time are therefore notched out and this gives a comb-like appearance to the resulting spectrum. With a single delay stage the notches are fairly wide, which gives an unpleasant 'drainpipe' colouration to the signal.

"However, by taking multiple delay stages in cascade, averaging their outputs, and subtracting this average signal from the original, a much sharper notch is obtained. This reduces the drainpipe effect and with five or six stages the signal is quite pleasant to listen to. The downside is that with the much sharper notches, it becomes necessary to frequently adjust the filter stage delay in order to track the wandering frequency of the mains supply. This implementation uses a simple feedback loop to maintain track of the mains period.

"The tracking works by running a couple of 'trial' filters at periods adjacent to the real filter period. If either of these filters does a better job of removing hum, the tracking algorithm adjusts the real period towards that of the better performing trial.

"The output from the `comb_sum` function provides a convenient way to determine which of the three trial periods offers the best filter. The function is called three times, once for each period, and each of the outputs is averaged. The trial with delay period closest to that of the incoming hum will give the highest average output because the input signal will be less coherent in the other two delay periods.

"The tracking algorithm needs to decide how far and how quickly to change the filter period in response to the varying outputs of its trial filters. It also needs to decide at what delay periods the trial filters should run, ie how far are they to be displaced from the 'real' period. The tracking will try to move these trial filters around as necessary for best results, so really there are three filters being tuned, not just the one!

"Some fairly ad hoc heuristics are employed and there is plenty of room for the experimenter to improve its performance.

"Sometimes the tracking can be upset by impulse noise, and this effect can be reduced by integrating the three `comb_sum` outputs, as shown in the example programme. This de-emphasises the higher-frequency signal components and can make for more stable tracking. However, if the input signal consists of mainly higher-frequency hum components (eg the signal has already passed through some high-pass filter) then it might be necessary to disable the integration of the `comb_sum` outputs in order to obtain reliable tracking.

"The delay stages are constructed from an array of signed 16-bit integers arranged as a circular buffer. The length of the buffer is rounded up to a power of two so that we can just use a bit-wise AND to cycle the pointers.

"Linear interpolation is used in the `comb_sum` function. This enables the delay to be adjusted to a fraction of a sample period, and makes for a fairly good notch even when the delay period is not spanned by an integer number of samples. More sophisticated interpolation might improve performance, but the simple linear interpolation is fast and should be adequate for most applications.

"Caveats and tips in no particular order:

- Delay tracking doesn't work for less than two stages.

- This type of filtering will only remove harmonic components of the foreground hum. Non-harmonic components can be a real problem, especially with signals from H-field antennas. Other techniques must be used to deal with these.

- It is vital to ensure that foreground hum does not clip anywhere in the receive path upstream of the sound card, and that the sound card mixer gain setting is such that the signal does not go out of range on the A/D. If overloading is allowed to occur, the hum cross-modulates with the noise to produce a broad spectrum of interference which cannot be removed.

- When dealing with foreground hum, 16-bit mode is preferable. If using eight-bit mode and the hum is 10 times louder than the signals, the remaining dynamic range after filtering is only five bits worth.

- In the event that the input hum is so weak that the filter can't recognise it, it will probably slew to one of its end stops and sit there harmlessly.

- When running with 10 or more stages, the filter can lock onto one of the dips in the pass-band ripple. This occurs if the filter is started from cold and the line frequency happens to be a long way from its nominal value. Usually after a few seconds the filter finds the correct notch.

- The ultimate depth of the notch is determined by the accuracy with which the outputs of the delay stages can be summed, which improves with increasing sample rate and increasing number of stages.

- There isn't much audible benefit in going beyond four or five stages."

I don't think it is necessary here to go step by step into the actual comb filter code and show it in a detailed fashion. Just a few points: the routine `SetupComb` is called during initialisation to allocate buffer memory and set some global variables. It is never called again. The actual filtering is performed by `ApplyComb`, to which are passed both the current input audio buffer composed of 512 audio samples, and the current setting of the fundamental frequency to be removed (50 or 60Hz). This because some parameters must be changed when this frequency changes. Filtering is done in-place, meaning that the contents of the input buffer are replaced by the filtered version.

A further note about the `leaky_integrate` function. It has been found that in some cases, the filter works better without this integration. It depends on the 'flavour' of the hum and also the amount of impulsive noise present too. It could be worthwhile to do some experiments disabling this function to check if the working of the filter is improved.

The function is easily disabled by replacing:

```
*a = (*a + val) * 0.999;
```

with simply:

```
*a = val;
```

## Conclusions

Well, we are the end of our journey into your first Windows programme. It is unlikely that you feel satisfied – either you will think that my description is too detailed or too brief. For those belonging to the first group, I can say that probably your knowledge of Windows is more advanced than that supposed for the intended audience of this chapter, and may be you wasted some of your time reading here. To the second group I can say that I sympathise with you, as I do remember quite well all of my problems when first trying to 'tame the beast', and was in search of all the information that it was possible to find. However, as I have already said a few pages back, going into much more detail would have meant to expand this chapter beyond its allotted limits, with the added risk of producing a poor replica of the Petzold book, which I recommend here once more.

In Table 10.14 on the following pages you will find the entire source code of Humid, which you will also be able to download, complete with an already compiled executable, from reference [3].

## References

[1] paul@abelian.demon.co.uk.
[2] www.borland.com/bcppbuilder/freecompiler/.
[3] http://www.qsl.net/i2phd/humid.zip.

### Table 10.14. Complete source code listing for HUMID

```
File : Humid.h
#include <windows.h>
#include <mmsystem.h>
#include <stdio.h>

  #define RATE       44100
  #define BUFSIZE    512
  #define OUTBUFS    512
  #define NINBUF     8
  #define NOUTBUF    32
  #define BYTEIN     2
  #define BYTEOUT    2
  #define ERRLEN     120
  #define NUM        (sizeof button / sizeof button[0])

static void ProcessBuffer(LPWAVEHDR ibuf);
static void fatal(char *);
static void Filter(void);
static LRESULT CALLBACK WndProc (HWND, UINT, WPARAM, LPARAM) ;

static BOOL SoundOpenInput(void);
static BOOL SoundOpenOutput(void);
static void SoundCloseInput(void);
static void SoundCloseOutput(void);
static void SoundStopInput(void);
static void SoundStopOutput(void);
static MMRESULT SoundStartInput(void);
static MMRESULT SoundStartOutput(void);

static VOID CALLBACK SoundProcIn   (HWAVEIN  , UINT , DWORD, DWORD, DWORD );
static VOID CALLBACK SoundProcOut  (HWAVEOUT , UINT , DWORD, DWORD, DWORD );
static DWORD  WINAPI ReQueBuffers(LPVOID);
static DWORD  WINAPI OutputAudio(LPVOID);

extern void ApplyComb(short*, float);
extern int  SetupComb(void);

static LPWAVEHDR GetFreeOutBuffer(void);




File : Humid.c
//------------------------------------------------
//   HUMID.C - Hum Instant Destroyer
//   Alberto di Bene I2PHD,  i2phd@weaksignals.com
//
//   Comb Filter code borrowed from
//   Paul Nicholson, paul@abelian.demon.co.uk
//------------------------------------------------

#include "humid.h"

struct
```

**Table 10.14 *(continued)***

```c
{
    long style ;
    char *text ;
    BOOL enabled;
}   button[] =
    {
        BS_PUSHBUTTON,      "50 Hz",    FALSE,
        BS_PUSHBUTTON,      "60 Hz",    TRUE,
        BS_PUSHBUTTON,      "Activate", FALSE,
        BS_PUSHBUTTON,      "Bypass",   TRUE
    };

BOOL       Hz50 = TRUE, Active = TRUE;
int        srate = RATE;
HWND       hwnd;
HANDLE     BufferThread, ProcessThread, InBufferReady, OutBufferReady;
HWAVEIN    hin;
HWAVEOUT   hout;
WAVEHDR    inhdr[NINBUF], outhdr[NOUTBUF];
RECT       rect, screen ;
//--------------------------------------
int WINAPI WinMain (HINSTANCE hInstance, HINSTANCE hPrevInstance,
                    PSTR szCmdLine, int iCmdShow)
{
    static char szAppName[] = "Humid" ;
    MSG         msg ;
    WNDCLASSEX  wndclass ;

    GetWindowRect(GetDesktopWindow(), &screen);

    rect.left = screen.right/2 - 250;
    rect.top  = screen.bottom/2 - 60;
    rect.right = rect.left + 500;
    rect.bottom = rect.top + 120;

    wndclass.cbSize        = sizeof (wndclass) ;
    wndclass.style         = CS_HREDRAW | CS_VREDRAW ;
    wndclass.lpfnWndProc   = WndProc ;
    wndclass.cbClsExtra    = 0 ;
    wndclass.cbWndExtra    = 0 ;
    wndclass.hInstance     = hInstance ;
    wndclass.hIcon         = LoadIcon (NULL, IDI_APPLICATION) ;
    wndclass.hCursor       = LoadCursor (NULL, IDC_HAND) ;
    wndclass.hbrBackground = (HBRUSH) GetStockObject (WHITE_BRUSH) ;
    wndclass.lpszMenuName  = NULL ;
    wndclass.lpszClassName = szAppName ;
    wndclass.hIconSm       = LoadIcon (NULL, IDI_APPLICATION) ;

    RegisterClassEx (&wndclass) ;

    hwnd = CreateWindow (szAppName, "  HUMID - Hum Instant Destroyer V1.2",
                         WS_CAPTION | WS_SYSMENU,
                         rect.left, rect.top,
                         500, 120,
```

**Table 10.14 *(continued)***

```
                        NULL, NULL, hInstance, NULL) ;

    ShowWindow (hwnd, iCmdShow) ;
    UpdateWindow (hwnd) ;

    while (GetMessage (&msg, NULL, 0, 0))
    {
        TranslateMessage (&msg) ;
        DispatchMessage (&msg) ;
    }


    return msg.wParam ;
}


//------------------------------------------------
LRESULT CALLBACK WndProc (HWND hwnd, UINT iMsg, WPARAM wParam, LPARAM lParam)
    {
    static char   msg[128] = "HUMID is now in filter mode, tuned at 50 Hz";
    static HWND   hwndButton[NUM] ;
    static int    cxChar, cyChar ;
    HDC           hdc ;
    PAINTSTRUCT   ps ;
    int           i, k;
    TEXTMETRIC    tm ;
    long          style;
    DWORD         thid1, thid2;

    switch (iMsg)
    {
        case WM_CREATE :
            hdc = GetDC (hwnd) ;
            SelectObject (hdc, GetStockObject (SYSTEM_FIXED_FONT)) ;
            GetTextMetrics (hdc, &tm) ;
            cxChar = tm.tmAveCharWidth ;
            cyChar = tm.tmHeight + tm.tmExternalLeading ;
            ReleaseDC (hwnd, hdc) ;

            for (i = 0 ; i < NUM ; i++)
            {
                style = WS_CHILD | WS_VISIBLE | button[i].style;
                if(!button[i].enabled) style |= WS_DISABLED;

                hwndButton[i] = CreateWindow ("button", button[i].text,
                            style,
                            cxChar * 15 * i + 25, cyChar * 3,
                            10 * cxChar, 7 * cyChar / 4,
                            hwnd, (HMENU) i,
                            ((LPCREATESTRUCT) lParam) -> hInstance, NULL) ;
            }

            if(!SetupComb())   // initialize the comb filter
            {
                fatal("Unable to allocate filter buffers");
                PostQuitMessage (1) ;
```

**Table 10.14 *(continued)***

```
        }


        if(!SoundOpenInput() || !SoundOpenOutput())
             PostQuitMessage (1) ;

        InBufferReady  = CreateEvent(NULL, FALSE, FALSE, NULL);
        OutBufferReady = CreateEvent(NULL, FALSE, FALSE, NULL);

        BufferThread   = CreateThread(NULL, 0, ReQueBuffers,
                                      0, 0, &thid1);
        ProcessThread  = CreateThread(NULL, 0, OutputAudio,
                                      0, 0, &thid2);
        if(NULL == BufferThread || NULL == ProcessThread)
        {
            fatal("Unable to create processing threads");
            PostQuitMessage (1) ;
        }

        SetThreadPriority(BufferThread,  THREAD_PRIORITY_HIGHEST);
        SetThreadPriority(ProcessThread, THREAD_PRIORITY_HIGHEST);

// if everything ok so far, then tell MMAPI to start recording
// but only when we will have plenty of free output buffers

        while(TRUE)
        {   for(k=0; k<NOUTBUF; k++)
                if(outhdr[k].dwUser != 0) break;

            if(k>=NOUTBUF/2) break;
        }
        waveInStart(hin);

        return 0;

    case WM_COMMAND :

        i = LOWORD (wParam);
        k = (i > 1) ? (5 - i) : (1 - i);

        button[i].enabled = FALSE;
        button[k].enabled = TRUE;

        EnableWindow(hwndButton[i], FALSE);
        EnableWindow(hwndButton[k], TRUE);

        Hz50   = !button[0].enabled;
        Active = !button[2].enabled;

        wsprintf(msg, "HUMID is now in %s mode", Active ? "filter" :
                                                          "bypass");
        if(Active)
        { strcat(msg, ", tuned at ");
          strcat(msg, Hz50 ? "50 Hz" : "60 Hz");
```

**Table 10.14** *(continued)*

```
            }
            else strcat(msg, "                          ");

            hdc = GetDC (hwnd) ;
            SelectObject (hdc, GetStockObject (SYSTEM_FIXED_FONT)) ;
            TextOut (hdc, 4 * cxChar, cyChar, msg, strlen(msg)) ;
            ReleaseDC (hwnd, hdc) ;

            return 0;

        case WM_PAINT :

            hdc = BeginPaint (hwnd, &ps) ;
            SelectObject (hdc, GetStockObject (SYSTEM_FIXED_FONT)) ;
            TextOut (hdc, 4 * cxChar, cyChar, msg, strlen(msg)) ;
            EndPaint (hwnd, &ps) ;

            return 0 ;

        case WM_CLOSE :

            PostQuitMessage (0) ;
            return 0 ;
    }


    return DefWindowProc (hwnd, iMsg, wParam, lParam) ;
}
//-------------------------------------------------
//
//      Sound_IO
//
//-------------------------------------------------

UINT          nByteIn=BYTEIN, nByteOut=BYTEOUT;
short         *inptr;

LPWAVEHDR     newOutBuffer;

static int    nDevIn, nDevOut;

static short  inbuf[NINBUF][BUFSIZE], tmpbuf[BUFSIZE];
static short  outbuf[NOUTBUF][OUTBUFS];

static BOOL sndInOpen = FALSE, sndOutOpen = FALSE, OutClosing, InClosing;



//-------------------------------------------------
//
// These functions start and stop the sound device
//
//-------------------------------------------------

BOOL SoundOpenInput(void)
```

**Table 10.14 (continued)**

```
{
    MMRESULT ierr;
    char err[ERRLEN];

    InClosing = FALSE;

    ierr = SoundStartInput();
    if (ierr != MMSYSERR_NOERROR)
    {
        if (ierr == 9) return FALSE;
        waveInGetErrorText(ierr, (LPSTR) err, ERRLEN);
        MessageBox(NULL, err, "Audio input error",
                   MB_APPLMODAL | MB_OK | MB_ICONEXCLAMATION);
        SoundCloseInput();
        return FALSE;
    }

    return TRUE;
}
//------------------------------------------
void SoundCloseInput(void)
{
    SoundStopInput();
}


//------------------------------------------
//
// These functions control opening and closing of the MCI device
//
//------------------------------------------

MMRESULT SoundStartInput(void)
{
    int i;
    MMRESULT ierr;
    WAVEFORMATEX    wf;

// first check that we have at least *ONE* MCI device

    nDevIn = waveInGetNumDevs() - 1;
    if (nDevIn < 0) return 1;

    ZeroMemory(&wf,sizeof(wf));     // clear the structure with 0
    wf.wFormatTag = WAVE_FORMAT_PCM;
    wf.nChannels = 1;
    wf.nSamplesPerSec = srate;
    wf.wBitsPerSample = (WORD) (8 * nByteIn);
    wf.nBlockAlign = (WORD)((wf.nChannels * wf.wBitsPerSample) >> 3);
    wf.nAvgBytesPerSec = wf.nSamplesPerSec * wf.nBlockAlign;
    wf.cbSize = 0;

    ierr = waveInOpen((LPHWAVEIN)&hin, WAVE_MAPPER, (LPWAVEFORMATEX)&wf,
                      (DWORD)SoundProcIn, 0L, CALLBACK_FUNCTION);
```

**Table 10.14 (continued)**

```c
    switch(ierr)
    {
        case MMSYSERR_ALLOCATED :
            fatal("Windows reports that your SoundCard is already in use by another pro-
gram");
            return 9;

        case MMSYSERR_BADDEVICEID :
            fatal("Bad SoundCard Device");
            return 9;

        case MMSYSERR_NODRIVER :
            fatal("No SoundCard driver!");
            return 9;

        case MMSYSERR_NOMEM :
            fatal("Not Enough Memory for SoundCard");
            return 9;

        case WAVERR_BADFORMAT :
            fatal("Your SoundCard does not support this Wave Format");
            return 9;
    }

// Everything ok. Our device is opened and ready to start recording
// We pass all the available buffers to MMAPI

    for(i=0; i<NINBUF && ierr == MMSYSERR_NOERROR; i++)
    {
        ZeroMemory(&inhdr[i],sizeof(inhdr[i]));
        ZeroMemory(inbuf[i],BUFSIZE * nByteIn);
        inhdr[i].lpData = (char *) inbuf[i];
        inhdr[i].dwBufferLength = BUFSIZE * nByteIn;/* The Length must be in bytes*/
        inhdr[i].dwBytesRecorded = 0;
        inhdr[i].dwUser = 0;
        inhdr[i].dwFlags = 0;
        inhdr[i].dwLoops = 0;
        inhdr[i].lpNext = NULL;
        inhdr[i].reserved = 0;
//first prepare the waveheaders
        ierr = waveInPrepareHeader(hin, &inhdr[i], sizeof(WAVEHDR));
//second add the buffers to the MMAPI queue
        ierr += waveInAddBuffer(hin, &inhdr[i], sizeof(WAVEHDR));
    }

    if(ierr!=MMSYSERR_NOERROR)
    {
        if(sndInOpen) waveInClose(hin);
        fatal("Errors during In_Buffers prepare");
    }

    return ierr;
}
```

**Table 10.14** *(continued)*

```
//-------------------------------------------------
void SoundStopInput(void)
{
    int i;

    InClosing = TRUE; // set this, or the threads will try
                      // to write to an invalid handle!
    waveInStop(hin);
    waveInReset(hin);

    for(i=0; i<NINBUF;  i++)
        waveInUnprepareHeader(hin,  &inhdr[i],  sizeof(WAVEHDR));
    if(sndInOpen)  waveInClose(hin);
}


//-----------------------------------------
//
// This function is the input callback from the MM Subsystem
//
//-----------------------------------------

VOID CALLBACK SoundProcIn(HWAVEIN hwi, UINT iMsg, DWORD dwInst,
                          DWORD dwParam1, DWORD dwParam2)
{
    switch(iMsg)
    {
    case WIM_DATA:
        if(!InClosing)
        {
            ProcessBuffer((LPWAVEHDR) dwParam1);
        }
        break;

    case WIM_OPEN:  // Message sent when input device is opened
        sndInOpen = TRUE;
        break;

    case WIM_CLOSE: // Message sent when input device is closed
        sndInOpen = FALSE;
        break;
    }
}
//-----------------------------------------
BOOL SoundOpenOutput(void)
{
    MMRESULT ierr;
    char err[ERRLEN];

    OutClosing = FALSE;

    ierr = SoundStartOutput();
    if (ierr != MMSYSERR_NOERROR)
    {
        if (ierr == 9) return FALSE;
```

**Table 10.14** *(continued)*

```c
            waveOutGetErrorText(ierr, (LPSTR) err, ERRLEN);
            MessageBox(NULL, err, "Audio output error",
                        MB_APPLMODAL | MB_OK | MB_ICONEXCLAMATION);
            SoundCloseOutput();
            return FALSE;
        }

    return TRUE;
}
//------------------------------------------------
void SoundCloseOutput(void)
{
    SoundStopOutput();
}


//------------------------------------------------
MMRESULT SoundStartOutput(void)
{
    int i;
    MMRESULT ierr;
    WAVEFORMATEX    wf;
// first check that we have at least *ONE* MCI device

    nDevOut = waveOutGetNumDevs() - 1;
    if (nDevOut < 0) return 1;

    ZeroMemory(&wf,sizeof(wf));    /* clear the structure with 0*/
    wf.wFormatTag = WAVE_FORMAT_PCM;
    wf.nChannels = 1;
    wf.nSamplesPerSec = srate;
    wf.wBitsPerSample = (WORD) (8 * nByteOut);
    wf.nBlockAlign = (WORD)((wf.nChannels * wf.wBitsPerSample) >> 3);
    wf.nAvgBytesPerSec = wf.nSamplesPerSec*wf.nBlockAlign;
    wf.cbSize = 0;

    ierr = waveOutOpen((LPHWAVEOUT)&hout, WAVE_MAPPER, (LPWAVEFORMATEX)&wf,
                        (DWORD)SoundProcOut, 0L, CALLBACK_FUNCTION);

    switch(ierr)
    {
        case MMSYSERR_ALLOCATED :
            fatal(
            "Windows reports that your SoundCard is already in use by another program\r\
                Please check if you did set fullduplex ON");
            return 9;

        case MMSYSERR_BADDEVICEID :
            fatal("Bad SoundCard Device");
            return 9;

        case MMSYSERR_NODRIVER :
            fatal("No SoundCard driver!");
            return 9;
```

**Table 10.14** *(continued)*

```c
        case MMSYSERR_NOMEM :
          fatal("Not Enough Memory for SoundCard");
          return 9;

        case WAVERR_BADFORMAT :
          fatal("Your SoundCard does not support this Wave Format");
          return 9;
    }

//the device starts playback immediately not like the recording device
//which needs another command to start recording, so pause it initially.

  ierr = waveOutPause(hout);

  for(i=0; i<NOUTBUF && ierr == MMSYSERR_NOERROR; i++)
  {   ZeroMemory(&outhdr[i],sizeof(outhdr[i]));
      ZeroMemory(outbuf[i],OUTBUFS * nByteOut);
      outhdr[i].lpData = (char *) outbuf[i];
      outhdr[i].dwBufferLength = (DWORD) OUTBUFS * nByteOut;
      outhdr[i].dwBytesRecorded = 0L;
      outhdr[i].dwUser = -1;     // -1 indicates : buffer queued
      outhdr[i].dwFlags = 0L;
      outhdr[i].dwLoops = 0L;
      outhdr[i].lpNext = NULL;
      outhdr[i].reserved = 0L;
      ierr = waveOutPrepareHeader(hout, &outhdr[i], sizeof(WAVEHDR));
  }

// queue the buffers
  for(i=0; i<NOUTBUF && ierr == MMSYSERR_NOERROR; i++)
     ierr = waveOutWrite(hout, &outhdr[i], sizeof(WAVEHDR));

  if(ierr!=MMSYSERR_NOERROR)
  {
    if(sndOutOpen) waveOutClose(hout);
    fatal("Errors during Out_Buffers prepare");
    return 1;
  }

  ierr = waveOutRestart(hout);
  return ierr;
}
//----------------------------------------------
void SoundStopOutput(void)
{
    int i;

    OutClosing = TRUE;

    waveOutPause(hout);
    waveOutReset(hout);

    for(i=0; i<NOUTBUF;  i++)
        waveOutUnprepareHeader(hout,  &outhdr[i],  sizeof(WAVEHDR));
```

**Table 10.14** *(continued)*

```
        if(sndOutOpen)  waveOutClose(hout);
}


//--------------------------------------------------
//
// This function is the output callback from the MM Subsystem
//
//--------------------------------------------------
VOID CALLBACK SoundProcOut(HWAVEOUT hwo, UINT iMsg, DWORD dwInst,
                                        DWORD dwParam1, DWORD dwParam2)

{
        switch(iMsg)
        {
          case WOM_DONE:
            if (!OutClosing)
            {
                ((LPWAVEHDR) dwParam1)->dwUser = 0;
            }
            break;

          case WOM_OPEN:  // Message sent when output device is opened
            sndOutOpen = TRUE;
            break;

          case WOM_CLOSE: // Message sent when output device is closed
            sndOutOpen = FALSE;
            break;

        }
}
//--------------------------------------------------
void ProcessBuffer(LPWAVEHDR ibuf)
{
    memcpy(tmpbuf, ibuf->lpData, BUFSIZE * sizeof(short));
    SetEvent(InBufferReady);
    Filter();
    SetEvent(OutBufferReady);
}
//--------------------------------------------------
void fatal(char * msg)
{
    MessageBox(hwnd, msg,"Humid", MB_APPLMODAL | MB_OK | MB_ICONEXCLAMATION);
}
//--------------------------------------------------
//      This is the buffer requeing thread, which cannot be done inside
//                      the MMSYSTEM Callback procedure

DWORD  WINAPI ReQueBuffers(LPVOID param)
{
    int i;
    MMRESULT ierr;

    while (TRUE)
    {
```

**Table 10.14 *(continued)***

```
          WaitForSingleObject(InBufferReady, INFINITE);

          for(i=0, ierr=0; i<NINBUF; i++)
            if(inhdr[i].dwFlags & WHDR_DONE)
              {
                if(MMSYSERR_NOERROR !=
                      (ierr = waveInAddBuffer(hin, &inhdr[i], sizeof(WAVEHDR))))
                  break;
              }

          if(ierr != MMSYSERR_NOERROR)
          { char s[128];
            sprintf(s, "Error during inbuffer requeing, code %d", ierr);
            fatal(s);
          }
        }
}
//-------------------------------------------
//    This is the audio processing thread, which requeues the filtered buffer
//                              to the MMSYSTEM queue

DWORD  WINAPI OutputAudio(LPVOID param)
{
    int i;

    while (TRUE)
    {
      WaitForSingleObject(OutBufferReady, INFINITE);

      for(i=0; i<NOUTBUF; i++)
        if(outhdr[i].dwUser == 1)
          {
            waveOutWrite(hout, &outhdr[i], sizeof(WAVEHDR));
            outhdr[i].dwUser = -1;
          }
    }
}
//-------------------------------------------
void Filter(void)
{
    if(NULL == (newOutBuffer = GetFreeOutBuffer())) return;
    if(Active) ApplyComb(tmpbuf, Hz50 ? 50. : 60.);
    memcpy(newOutBuffer->lpData, tmpbuf, BUFSIZE * sizeof(short));
}
//-------------------------------------------
LPWAVEHDR GetFreeOutBuffer(void)
{
    int k;

    for(k=0; k<NOUTBUF; k++)
      if(outhdr[k].dwUser == 0)
        {
          outhdr[k].dwUser = 1;
          return &outhdr[k];
```

**Table 10.14** *(continued)*

```
      }

   return NULL;
}
//---------------------------------------------


File : CombFilter.h
#define VERSION "1.2"

/*
 *  End stop.  Maximum +/- percentage deviation from the nominal delay that
 *  the tracking will be allowed to use.
 */
#define END_STOP 0.5     /* Percent */

/*
 *  Number of comb delay stages.
 */
#define DEFAULT_STAGES 4

/*
 *  Seconds between iterations of the line frequency tracking.
 */
#define DEFAULT_TRACKING_CYCLE 1

/*
 *  The default encoding, 8 means unsigned bytes, 16 means signed 16 bit words.
 */
#define DEFAULT_ENCODING 16

/*
 *  This constant sets the nominal slewing rate of the period tracking. This
 *  value gives stable locking under most conditions. The filter may lag when
 *  a sudden shift occurs, if this happens too much,  then raise the value,
 *  eg to 0.02 or 0.04.   If the filter jumps out of lock on bursts of noise,
 *  then reduce the value.
 */
#define SLEW_RATE 0.01

/*
 *  Smoothing factor for the period tracking.  Not very critical.
 */
#define COMB_SMOOTH 0.9999

/*
 *  Maximum tracking width.  The max delay difference between the low and high
 *  measurement periods, and the current delay period, units of samples.
 */
#define MAX_TRACK 10

/*
 *  End of tunable definitions.
 */
```

**Table 10.14 (continued)**

```c
extern BOOL Hz50;

File : CombFilter.c
#include "Humid.h"
#include "CombFilter.h"
#include <math.h>

//*************************************************************************
//          The Comb Filter
//*************************************************************************

static short *comb_buffer;     // Points to comb delay buffer
static int    comb_len;        // Total length of the comb delay buffer
static int    comb_lp = 0;     // Load offset into comb_buffer, next location to load
static int    comb_mask;       // Mask for buffer offsets
static int    comb_stages = DEFAULT_STAGES;   // The number of delay stages in use

static float comb_delay;       // Current comb delay, per stage, in units of samples
static float comb_adj;         // Incremental adjustment to comb delay
static float max_delay, min_delay;        // Allowed limits to comb delay

static float track_low  = 0;   // Low period accumulator
static float track_mid  = 0;   // Middle period accumulator - current delay
static float track_high = 0;   // High period accumulator

static float tracking_width = 1.0; // Offset in period space of the tracking guides
static float tracking_cycle = DEFAULT_TRACKING_CYCLE;   // How often to revise delay
static float nom_delay;         // nominal delay : 20 ms for 50 Hz, 16.666 for 60 Hz


//------------------------------------------
/*
 * Initialise the comb filter: allocate buffer memory and set some globals.
 * The buffer must be long enough for the max possible delay, plus an extra
 * allowance for the high period tracking. We then round up to the next
 * power of two for efficiency.
 */

int SetupComb(void)
{
   int i, max_comb_len;

   max_comb_len = 1 + MAX_TRACK * comb_stages +
                 20.0 * comb_stages * RATE/1000. * (1+END_STOP/100.0);

   for(i=1; i < max_comb_len;) i <<= 1;

   max_comb_len = i;

   if((comb_buffer = (short *) malloc( sizeof(short) * max_comb_len)) == NULL)
   {
      return 0;
   }
```

**Table 10.14** *(continued)*

```
   return 1;
}

//-------------------------------------------------
//
//  Comb summing function.  Sum over the given number of delay stages, each
//  of length delay_steps samples.  Use linear interpolation between samples.
//
//  At high sample rates, adequate results can be obtained without the linear
//  interpolation, which can save quite a bit of CPU time.
//

static int __inline comb_sum(int stages, float delay_steps)
{
   int ne;
   float f = 0, g = delay_steps;

   for( ne=1; ne<=stages; ne++, g += delay_steps)
   {
      int gf = g;
      int p, f1, f2;

      p = comb_lp - gf;
      f1 = comb_buffer[ p- & comb_mask];
      f2 = comb_buffer[ p & comb_mask];
      f += f1 + (f2-f1) * (g-gf);
   }

   return (int)(f/stages + 0.5);
}


//-------------------------------------------------

static void adjust_delay(void)
{
   float f;
   float balance = track_high/track_low;    /* Equals 1.0 when in tune */
   float mean = track_high * track_low/track_mid/track_mid;

   if( balance > 1) comb_adj = 1;
   else
   {
      comb_adj = -1;
      balance = 1/balance;
   }

   f = pow( balance, 10);  if(f > 10) f = 10;
   comb_adj *= f * SLEW_RATE/comb_stages;

   if( mean < 0.5) tracking_width /= 2;
   else
   if( mean > 0.8) tracking_width *= 2;
```

**Table 10.14 (continued)**

```c
   if( tracking_width < 0.1) tracking_width = 0.1;
   else
   if( tracking_width > MAX_TRACK) tracking_width = MAX_TRACK;

// Revise the delay, and constrain within the end stops.

   comb_delay += comb_adj * RATE/1000.0;
   if( comb_delay > max_delay) comb_delay = max_delay;
   if( comb_delay < min_delay) comb_delay = min_delay;
}


//----------------------------------------------
static void __inline smooth_accumulate(float *a, float val)
{
   *a = *a * COMB_SMOOTH + val * (1-COMB_SMOOTH);
}
//----------------------------------------------
static void __inline leaky_integrate( float *a, float val)
{
   *a = (*a + val) * 0.999;
}


//----------------------------------------------
//         This function is called with each signal buffer

void ApplyComb(short* inbuf, float freq)
{
   static float p_low = 0., p_mid = 0., p_high = 0.;
   static float oldfreq = 0.;
   static int count = 0;
   int i, t, fg, idx;
   short  s;

   if(freq != oldfreq)
   {
     oldfreq    = freq;
     nom_delay  = 1000./freq;
     comb_delay = nom_delay * RATE/1000.0;
     max_delay = comb_delay * (1+END_STOP/100.0);
     min_delay = comb_delay * (1-END_STOP/100.0);

     comb_len = comb_stages * (1 + MAX_TRACK + max_delay);

     for( i=1; i < comb_len; ) i <<= 1;

     comb_len = i;
     comb_mask = comb_len-1;
   }

// Process the incoming buffer, and keep track of the mean
// square and peak values.

   for(idx=0; idx<BUFSIZE; idx++)
```

**Table 10.14** *(continued)*

```
  {
    s = inbuf[idx];
    comb_buffer[comb_lp] = s;    // Store incoming sample in comb buffer

// Run the comb summing function three times, for the low, high, and middle
// periods.  Integrate each sum over time to give a 6db/octave roll-off.
// Then smooth the squared amplitudes of the integrated sums.

    t = comb_sum( comb_stages, comb_delay - tracking_width);
    leaky_integrate( &p_low, t);
    smooth_accumulate( &track_low, p_low * p_low);

    t = comb_sum( comb_stages, comb_delay + tracking_width);
    leaky_integrate( &p_high, t);
    smooth_accumulate( &track_high, p_high * p_high);

    t = comb_sum( comb_stages, comb_delay);
    leaky_integrate( &p_mid, t);
    smooth_accumulate( &track_mid, p_mid * p_mid);

// Revise the delay every tracking_cycle secs.

    if( ++count > RATE * tracking_cycle)
    {
       if( track_mid != 0)    // Must have a non-zero input signal
       {
          adjust_delay();
       }
    }

    comb_lp = (comb_lp+1) & comb_mask;
    inbuf[idx] = s - t;
  }
}
//--------------------------------------------
```

# Index

Extra resources for this book are at www.rsgb.org/books/extra/command.htm